Deskbound

Standing Up to a Sitting World

Dr. Kelly Starrett

with Juliet Starrett and Glen Cordoza

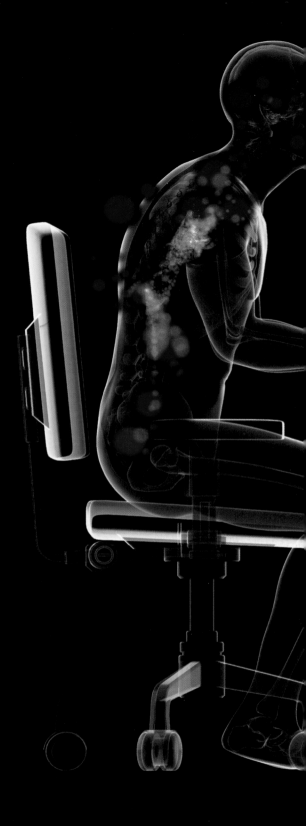

VICTORY BELT PUBLISHING

Las Vegas

First published in 2016 by Victory Belt Publishing Inc.

ISBN-13: 978-1628600-58-2

Cover Design: Tom Wiscombe, TomWiscombe.com

Design: Yordan Terziev and Boryana Yordanova

Printed in the USA
RRD 0116

This book is dedicated to Principal Tracy Smith and the students and staff at Vallecito Elementary School in San Rafael, California—the first all-standing school in the world.

Table of Contents

Introduction

You've already heard that sitting is the new smoking. While this may sound like ridiculously overblown media hype, the author of the phrase isn't backing down. He is Dr. James Levine, director of the Mayo Clinic–Arizona State University Obesity Solutions Initiative, and he'll even go further. He says, "Sitting is more dangerous than smoking, kills more people than HIV and is more treacherous than parachuting." He concludes simply, "We are sitting ourselves to death."[1]

The good doctor isn't the lone alarmist. Dr. Levine and a swiftly growing number of experts, backed by a mountain of research, argue that sitting for as little as two continuous hours increases the risk of heart disease, diabetes, metabolic syndrome, cancer, back and neck pain, and other orthopedic problems.[2] Sitting will shorten your life, just like smoking.

Many studies are also demonstrating that the effects of long-term sitting aren't reversible through exercise or other good habits. This means that if you eat well and work out religiously for an hour a day, but then sit for all or most of your other waking hours, the sitting behavior will chip away or even cancel out the benefits of all your exercise at the gym.[3] You are still considered a sedentary person.

Some experts argue that sitting is even more pernicious than smoking. An Australian study conducted in 2008 reports that every hour of television watched after age 25 reduces the viewer's life expectancy by 21.8 minutes.[4] By comparison, smoking a single cigarette reduces life expectancy by 11 minutes.[5] Dr. Levine claims that for every hour we sit, we lose two hours of life.[6]

The typical seated office worker has more musculoskeletal injuries than any other industry sector worker, including construction, metal industry, and transportation workers. One researcher's conclusion: sitting is as much an occupational risk as lifting heavy weights on the job.[7]

For the past two decades, doctors and research scientists have been studying the deadly impact of sitting too much. The media has more recently caught up to name the issue a public health crisis because of the broad body of evidence linking sedentary behavior to a wide range of negative health outcomes. Today, the World Health Organization ranks physical inactivity—sitting too much—as the fourth biggest preventable killer globally, causing an estimated 3.2 million

The Ramifications of Sitting

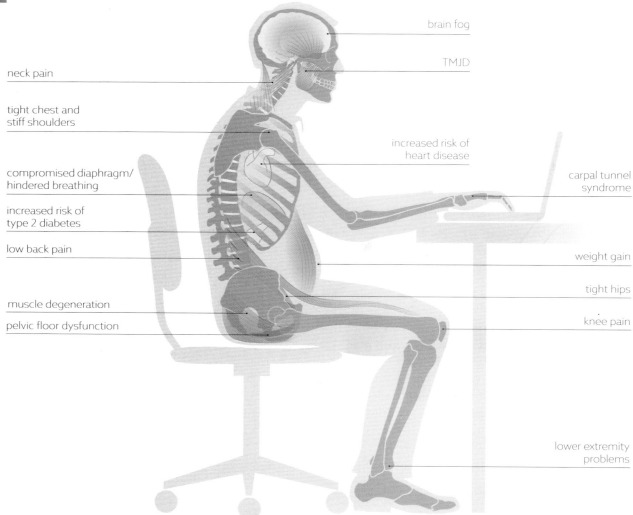

brain fog

TMJD

neck pain

tight chest and
stiff shoulders

increased risk of
heart disease

compromised diaphragm/
hindered breathing

carpal tunnel
syndrome

increased risk of
type 2 diabetes

low back pain

weight gain

tight hips

muscle degeneration

knee pain

pelvic floor dysfunction

lower extremity
problems

SITTING
while eating breakfast

SITTING
on the way to work

SITTING
at work

SITTING
in front of a computer

deaths annually.[8] In just the last 20 years, the simple act of sitting has leapfrogged to the top of the health killer charts globally.

How do we make sense of this—and fix it? The answer is simple. Unless we are asleep, we human beings are designed to move. Our normal physiology depends on this important fact. We evolved a central nervous system to sense changes around us and to move us through our environment. For nearly 200,000 years, *Homo sapiens* spent the majority of their time on the move. If they wanted something to eat, they had to hunt for it or dig it up out of the ground. If they wanted to travel, they had to get there by foot. All the movement that surrounded survival might sound exhausting to us now, but that activity shaped our bodies' design, both internally and externally. Our bodies are built for movement, and in turn movement keeps our bodies healthy. It is a symbiotic relationship that has allowed our species to survive.

The problem with evolution is that it has no foresight. It couldn't predict the invention of the chair. In the beginning, this simple four-legged piece of furniture didn't have a notable impact on human health. For most people, it was little more than a place to rest after a hard day in the field or factory. Fast-forward to the 21st century. In an astonishingly short period, the citizens of the developed nations of the world have become almost completely sedentary—from the way we shop and travel to the way we work and play. Today, Americans spend an average of 13 hours a day sitting.[9]

Once the desk/chair combination became the cultural norm in the workplace, other sitting-based innovations followed. The intercom allowed office workers to communicate without getting up from their chairs. The TV set seduced people of all ages into passive leisure-time activity. In the 1950s, when cars had become affordable to the general public and the interstate highway system was developed, people began flocking to the suburbs, and the commuter culture was kick-started. Then, of course, came the computer, and our fate as sedentary creatures was sealed. We had become deskbound.

The Surgeon General's first report on physical inactivity and health was published in 1996. Similar to the Surgeon General's 1964 report on tobacco, it illuminated the broad body of evidence linking sedentary behavior to a wide range of negative health outcomes.[10]

SITTING
while eating dinner

SITTING
while slouched over a phone

SITTING
in front of the TV

The problem with sitting is that it seems so innocent and natural. Our bodies easily bend into that shape, so how could it be bad for us? Of course, if we sat with perfect posture for 15 minutes and then spent the rest of our waking hours moving, then that short stint of sitting wouldn't be bad for us. But sitting is like eating potato chips—rarely do we do it in moderation.

When we sit for long periods, the muscles in our lower bodies literally turn off and become inactive. Simultaneously, we automatically adopt positions that don't utilize the critical muscles and connective tissues that stabilize and support our trunk and spine. The result is compromised body function, and it causes a multitude of common and pernicious orthopedic problems, like back and neck dysfunction, carpal tunnel syndrome, and pelvic floor dysfunction. When we travel around the world to teach and consult, we always start by asking the audience members to raise their hands if they are *not* in pain. Invariably, only 5 to 10 percent of every room—and this includes diverse groups like corporate office workers, members of the military, professional athletes, and even children—raise their hands. This means that 90 to 95 percent of the people we work with are in pain. We believe that one of the root causes of this widespread self-reported pain is too much sitting.

Sitting too much is not just shortening our lives and causing us pain, it's costing us. The Centers for Disease Control reports that we are spending 75 cents of every health-care dollar on chronic conditions linked to sedentary behavior, like obesity, diabetes, and heart disease.[11] According to the National Institutes of Heath, back pain affects 8 of 10 people in their lifetime, and it is the leading cause of disability worldwide.[12] In the United States alone, we spend almost $1 billion treating back pain[13] and $20 billion in employer costs treating carpal tunnel[14] annually. The indirect costs are even higher once we consider lost workdays and decreased productivity. And these figures are just the tip of the iceberg in terms of what we are spending to fix our ailing sedentary bodies.

If you're like most people, you might find this somewhat hard to believe. You might suspect that we are blowing things out of proportion. But the evidence continues to mount that sitting too much is terrible for us. If our brains worked differently and we could vividly recall the normal feelings of childhood—running, jumping, and crawling without pain or restriction—then we might view what we as adults consider "normal" with more disdain. We might go searching a little harder for the cause of our suffering.

We don't, though, because in our sedentary world, the actual cause of pain or sickness can be very elusive, and assigning blame can be difficult. If a beaver chews away at a tree for nine days straight and then a mild breeze knocks the tree over, what caused that tree to fall? Sure, the wind finished the job, but without the beaver's long hours of arduous work, the tree would have been just fine. When it comes to many modern ailments, our sedentary lifestyle is the hardworking beaver that weakens our bodies and primes us for pain and disease. It's time we stand up to our sitting world.

The Elephant in the Room

Early on in Kelly's career as a physical therapist and strength and conditioning coach, he came to understand that in order to be effective at his job, he needed to play the role of observer. The first thing he noticed was that an unusual number of people were in pain. Operating under the assumption that our joints are designed to last 110 years and that a human being's natural resting state is pain-free, he was dumbfounded as to why so many people were in utter agony in their thirties and even their twenties.

To get to the bottom of this issue, we needed to analyze our modern lifestyle as a whole, figure out what was causing the greatest harm to our bodies, and then find unobtrusive ways to mitigate those noxious influences. Well, when we opened a strength and conditioning center in 2005, we had the chance to observe a great many people. Since the doors opened, we've worked with the NFL, NBA, MLB, and NHL, dozens of Division I university programs, MMA fighters, extreme athletes, ballet dancers, elite cyclists, Olympians, every branch of the military, including their elite soldiers, and even A-list Hollywood actors. We've also worked with weekend warriors, desk jockeys, and children. While our job was to focus on individual clients and their unique pain, we started noticing a common thread. Even after we corrected mechanical (technique) problems and/or range of motion (flexibility) restrictions, many of our clients continued to have problems excelling at their sport or job.

In 2007, we were consulting with a Division I football program. We had been asked to help them figure out why, despite a focus on proper training and injury prevention, they were still seeing an alarming injury rate and lost power output. The athletes were also reporting a high rate of low back and lower extremity pain, which was alarming to the coaches. We evaluated the program and found that they were doing all the right things—they had an excellent strength and conditioning program and a keen understanding of the importance of mechanics and range of motion, their athletes were dedicated and hardworking, and they had an excellent and thoughtful coaching staff. When we dug deeper, we discovered that outside of practice, the athletes were spending 12 to 14 hours a day sitting. Let us repeat—Division 1 college football players were sitting for almost all of their waking hours, except when they were actually playing or practicing! They sat in class, sat during team meetings, sat in the cafeteria, sat on the couch, and so on. All of this sitting led to two problems: it left very little time for physical activity (movement), and the athletes were stuck in compromised positions for the majority of the day.

We had discovered our elephant in the room.

We started calling this phenomenon "innocuous environmental load." We use this term to describe seemingly harmless aspects of the environment that place compromising stressors on the body and can hinder physiology and basic life functionality. For example, if you have a newborn baby at home, you're probably not sleeping very well. This is an environmental load, or stressor, that is impacting the quality of your existence. Lack of sleep is an easy one to diagnose, because those clients who suffer from it have bags under their eyes and often proclaim, "The baby kept me up all night!" The solution is simple—get more sleep and you will perform better. But not all environmental loads are so easy to pinpoint.

Digging deeper and deeper into our clients' lives outside the gym, we started to realize that those who were hitting the biggest roadblocks had a unifying thread: they all spent a great deal of time sitting. Even when we corrected their mechanics in the gym, over and over again we had to solve the same set of physical problems related to sitting—tight hips, low back and neck pain, shoulder restrictions . . . the list goes on and on. It would have made more sense if these were prototypical unhealthy people, but many of them were top-notch athletes who trained regularly, slept an adequate amount, managed their life stress, and ate well. Their bodies should have been well-oiled machines, yet they couldn't overcome the environmental load of sitting.

The true impact on the body that sitting can have became even more evident to us when we began working with elite military aircraft pilots. Though they were well-trained tactical athletes, their bodies were a mess. Think about what these pilots go through on a typical day: in addition to facing high-stress situations from a seated position, they experience insane g-forces, oscillate up and down, and wear heavy helmets, and the seat forces their shoulders forward into an ergonomically poor rounded position. Just as with our clients in the gym, we saw all the familiar problems that go along with prolonged sitting, but intensified and accelerated—disc and neck herniation, numbness and tingling, chronic back pain. And when they crawled out of their planes and tried to get on with the other aspects of their lives, not only did they suffer from pain, but their mobility and performance also were affected.

Our bodies are designed to handle serious abuse, but like everything else, the human body has its limits. And at every turn our chair-centric culture pushes those limits, which compels us to ask a very important question . . .

Where Did We Go Wrong?

Here is a wonderful fact: your body adapts to the position that you assume for most of the day. So, if you sit and allow your back to round forward (flexion) or arch backward (overextension), your tissues and joints will form a sort of cast around that posture, making it difficult for you to get into better positions later.

What most people fail to realize is that the positions we assume for most of the day also impact the ways we move the rest of the day. And the quality of our movement (or lack thereof) can affect the quality of our lives. But at what point in life do innocuous environmental loads start to creep in and compromise mechanics? In our experience, the adaptation starts showing up around the first grade.

As part of the research for his book *Ready to Run,* Kelly began observing the kids at our daughter's school as they ran. Every child in her kindergarten class was a beautiful runner. Their running was mechanically sound and technically perfect. It didn't matter whether they ran fast or slow, barefoot or with shoes, they all ran in the same natural way, like little Olympic sprinters. About halfway through the first grade, however, Kelly noticed that half of the kids had started to heel-strike—landing on their heels instead of the balls of their feet. Literally half of them had begun to adopt a brand-new movement pattern that was fundamentally different from the one they had used since they started running. (As an experiment, watch how many runners in the next Olympics slam their heels into the ground like brakes as they run. The answer will be none.) When he observed this, he immediately asked himself, "What the heck happened in the last year?" The children instinctively knew how to run. No one had taken them aside and instructed them on how to position their feet when landing; they just knew how to do it because it came naturally. But sometime during first grade, they began running in a way that we unequivocally know leads to problems down the road. Seriously, what the heck happened?

We realized that the innocuous environmental load for these children was sitting all day at school. Because kindergarten is generally more active than other grades, kids don't really begin their sitting career until first grade. The effects are almost immediate, in the form of primary movement mechanics changing to dysfunctional heel-striking running mechanics. In just a short period, we saw that the kids had adapted to their new environment, and heel-strike running was that adaptation. The new movement pattern was the expression of what was happening upstream at the tissue level. Heel-strike running isn't what we humans evolved to do. It doesn't allow us to use our springy heel cords, for example, and it can only be done while wearing highly cushioned shoes. When a soldier loses a lower limb and is fitted with a prosthetic, he has to be retaught how to run naturally if he was a heel-striker. Why? Because in spite of the incredible advances in material science, we cannot engineer an artificial limb that can withstand the loading forces of heel-striking.

Heel-striking is a problem, and it shouldn't exist in the running human. Tell us, can you think of any animal that adopts multiple running patterns?

A 2010 American Cancer Society study that followed 123,216 adults for 13 years showed that women who were inactive and sat for more than six hours a day were 94 percent more likely to die during the period studied than those who were physically active and sat for less than three hours a day. Men who were inactive and sat for more than six hours a day were 48 percent more likely to die than their active counterparts. Notably, these findings were independent of physical activity levels—the negative effects of sitting were just as strong in people who exercised regularly.[15]

Children
and Sitting

Only 31 percent of California students passed all six areas assessed in a statewide physical fitness test in 2011, in part because of budget cuts to physical education programs. In a 2011 survey released by the California State PTA, 75 percent of respondents reported that their children's physical education or sports programs had been cut or reduced dramatically.[16]

The U.S. Department of Health and Human Services' Physical Activity Guidelines for Americans recommend that children get at least 60 minutes of daily moderate to vigorous physical activity.[17] Most children today get far, far less than that. It is estimated that only 4 percent of elementary schools, 8 percent of middle schools, and 2 percent of high schools provide daily physical education.[18]

A groundbreaking survey conducted by the Kaiser Family Foundation showed that children between ages 8 and 18 spend an average of 7.5 hours a day sitting in front of a screen, regardless of socioeconomic status.[19] Add to that the 4 to 6 hours that children spend at their desks at school, sitting in cars, eating meals, and doing homework. Our kids are spending 10 to 14 hours a day—a whopping 85 percent of their waking hours—in sedentary pursuits.[20] Only one in four children ages 6 to 15 meets the 2008 Physical Activity Guidelines for Americans recommendation of at least 60 minutes of moderate to vigorous physical activity per day.[21]

The Centers for Disease Control reports that only 13 percent of children walk to school today, compared with 66 percent in 1970.[22] Among students living within 1 mile of school, the number who walk plummeted from 90 percent to 31 percent between 1969 and 2001.[23]

In 1980, there were 81 million TVs in homes. Today, the number has more than tripled, to about 324 million.[24] During the same period, the number of obese children and adolescents also tripled.[25] According to childhood obesity researchers at the University of California at Berkeley, for every hour that children ages 12 to 17 sit watching TV, their likelihood of being overweight increases by 2 percent.[26]

In March 2004, U.S. Surgeon General Richard Carmona warned that because of swiftly rising rates of obesity, unhealthy eating habits, and physical inactivity, "we may see the first generation that will be less healthy and have a shorter life expectancy than their parents."[27]

If you think that the statistics don't apply to you because your child participates in an after-school sport, we've got some bad news. Unfortunately, when exercise is tacked onto an otherwise sedentary lifestyle—in both children and adults—all those pitches and field goals don't counteract the negative effects of too much sitting.[28] Too much sitting outside of a soccer or baseball game is too much sitting. Period.

New research shows that continued utilization of standing desks was associated with significant improvements in executive function and working memory capabilities in schoolchildren.[29]

You can't, because all animals run the way their physiology intended. If they don't, they get injured and wash out of the gene pool. Eighty percent of the runners in America are injured every year,[30] making running one of the most dangerous activities in which we engage. And here's the rub: just as we take the changes in the way children may run as normal, we have accepted that human beings, who were designed and evolved to run long distances over many decades, are going to get injured while running. The problem remains because human mechanics are so robust that we can't always see the implications of moving poorly right away. They can be observed only when we stretch the timeline. And this is a problem.

At about this same time, we witnessed that more and more of the first graders were adopting rounded-forward, flexed-spinal positions. Every time we dropped into a classroom, we found that nearly every child was sitting with a horrible shrimp back. (You can perform this same experiment at your local coffee shop. Take an espresso shot for any person you see sitting with a C-shaped spine. Sorry in advance about the jitters.) And it was clear that sitting was also patterning poor spinal mechanics while moving.

With each passing grade, the problems got worse. The kids' instincts about how to run and move naturally were slowly fading away, and the burdens of modern life were starting to create a new normal for their bodies, until moving and running the way they did in kindergarten seemed foreign to them. If heel-striking or moving with a flexed spine had been painful, we're sure that all of them would have reverted to their natural patterns. But our bodies are designed to take a lot of abuse and constantly adapt to survive in the environment by expending the least amount of energy. Practice doesn't make perfect, it makes permanent, and the ramifications of ugly movement patterns often don't rear their heads until much later.

Studies show that the more you sit, the worse your cholesterol panel. Sitting too much also increases heart attack risk, causes arteries to stiffen, and results in skeletal softening. What's more, breast, colon, lung, and endometrial cancers are associated with too much sitting.[32]

When we say "much later," we wish we meant 30 or 40 years down the road, but as out of sync as we have become with our natural movement patterns, that often isn't the case. Classic orthopedic pathologies like neck and back pain are even showing up in teenagers. Dr. James Carter, an Australian chiropractor, recently made headlines when he showed x-ray images of children as young as seven who had developed abnormal spinal curves and bone abnormalities from spending too much time on smartphones and sitting too much.[31]

Once we realized the impact that sitting had on our children, clients, and athletes, we immediately developed a system to combat the problem. The solution that we devised is the same system that we cover in this book. It encompasses strategies for increasing movement (activity) and prioritizing mechanics (how to position and move your body correctly). This isn't an easy shift, however. Most of the people we work with already consider themselves active. What they fail to realize is that exercise, even intense exercise, and activity are two different things.

Why Exercise Isn't Enough

It would be so much easier if exercise could undo the damage caused by sitting. We could just pull a desk jockey into the gym and teach her how to properly perform a host of movements, and she would slowly make improvements without having to alter her actions outside of the gym. We wouldn't have to dive into her personal business and tell her that she should also pay attention to her body the other 23 hours of the day. But the cold, hard truth is that exercise will not reverse the potentially harmful and irrefutable effects that too much sitting has on our bodies.

This shouldn't come as too big of a surprise. We've known about the benefits of exercise for a long time, and we've also known that it's not a cure-all for poor lifestyle choices. We wouldn't consider someone who consumes nothing but fast food and soft drinks healthy, even if she regularly makes it to the gym. And we wouldn't consider someone healthy if he gets raging drunk seven days a week, even though he somehow musters the strength to go for a run every morning. Sure, exercise might help the body rid itself of some empty calories or get stronger, but most people are informed enough to realize that it's not going to magically overcome an unhealthy lifestyle.

For some reason, people don't seem to apply the same logic to the ravages of sitting. Some of the smartest people we know think that they can shake off eight or more hours of sitting with a hard workout. We don't mean to be rude, but this makes about as much sense as thinking that you can walk off a broken foot. When you sit for prolonged periods, your body is forced into compromised positions, which leads to compromised function. And when you sit all day, you simply don't move enough. Working out will certainly make you healthier overall, but it's not a time machine that can undo the sedentary choices that you make over the rest of the day.

Understanding this can help you make sense of some of the roadblocks you may be hitting with your health and well-being. For example, the training programs of many professional and recreational athletes are halted by movement problems and tissue dysfunction. In an attempt to find a solution, the athletes check and recheck their lifting form, see a host of physical therapists or other body workers, or even try new workout regimens. But no matter how organized they get their bodies during their workouts, they can't escape the ramifications of sitting too much. While the deterioration process may be slow, it is methodical, and if left unchecked, it eventually will have a huge impact on all aspects of their lives, including their resiliency and performance.

Another roadblock that many deskbound athletes encounter is weight gain. When they come to us for advice, we ask them to describe their lifestyle. They almost always describe themselves as being active, even though they have jobs that require them to sit at a desk for 8 to 10 hours a day. They tell

A British study of more than 4,000 civil servants found that spending less than 12 hours a week sitting decreased diabetes risk by 75 percent. Those who sat more than 25 hours a week had an increased chance of developing metabolic risk factors like diabetes, insulin resistance, and "bad" cholesterol.[33]

us how, as the weight comes on, they work out longer and harder, but with diminishing results. That's when we explain to them the difference between "exercise" and "non-exercise activity." This is critical.

"Exercise" typically means activities like running or weight training. Despite the intensity of the workout, you usually do it for a relatively short period: hit the gym or track, work out, and then head home. Though no one would argue that exercise isn't essential, working out by itself won't compensate for the problems associated with sedentary behavior. That's because a standing body uses energy quite differently than a sitting body. Depending on our level of activity, we burn calories at different rates, and we also store calories differently. "Non-exercise activity" is what you do when you're not exercising at high intensity—things like standing, walking, gardening, cooking, running to catch an elevator, and even fidgeting.

We're not making this up. Dr. James Levine calls it Non-Exercise Activity Thermogenesis, or NEAT.[34] According to Dr. Levine, someone who is deskbound for eight hours a day burns approximately 300 NEAT calories during that period. In comparison, someone who is not deskbound, such as a waiter, burns around 1,300 NEAT calories. That's a difference of 1,000 calories! Over time, that calorie variance can be the difference between being thin and being obese. And for someone who is deskbound and doesn't exercise, it could quickly lead to obesity and all the negative health consequences that go along with it.

A sitting body doesn't expend much energy. The signals that normally are sent as a result of moving and burning calories literally turn off. Simultaneously, the processes that store fat are summoned to duty. The result: people with low levels of non-exercise activity are predisposed to obesity. Obesity and sedentary behavior (sitting) go hand in hand.

A Blueprint for Human Movement

We're sure you've heard some iteration of the "sit less and stand more" mantra. While these efforts are certainly beneficial to our overall health, they leave a big piece of the problem on the table.

Having worked with many thousands of athletes of all shapes, sizes, and skill levels, and having seen how the negative impacts of sitting affect everyone, not just those who are inactive, it became very apparent to us that in our modern world, we need a blueprint for how to move.

Few of us have strenuous jobs that require nonstop NEAT movement. Instead, the vast majority of us are in front of a computer all day, sitting for most of our waking hours. And sitting in a chair doesn't reflect our need to move. What was once inherent has become a skill that we must learn. This used to be the job of physical education teachers—in the not-too-distant past, those teachers taught kids how to move. Kids learned how to squat and run and jump and land and climb ropes. Today, physical education in schools is about teaching kids sports skills, which are distinct from movement skills. We teach our children to read and write, do math, take care of their hygiene, and use technology, but the art of teaching movement skills is a thing of the past, and we have forgotten to value it. In order to remain healthy and pain-free as a society, we must view movement as something that needs to be taught, practiced, refined, and appreciated—and is as important as anything else we learn. Our health depends on it.

We know what you might be thinking: "But I put my child in gymnastics, and I take Pilates and yoga. Don't those things help?" Yes, they certainly do. Gymnastics can develop agility and strength, and yoga can improve flexibility and spinal mechanics. However, none of these systems offers a seamless transition to everyday movements like hefting a kid out of a crib or picking up a bag of groceries. We need a scalable system that teaches us how to move in all situations—from ordinary daily-life activities to the highest levels of performance. In other words, the system you use to get your body organized for sitting down in your office chair should be the same as the system you use to get your body organized for training in the gym.

If you're one of the few individuals who have been moving improperly your entire life and have suffered no ill consequences, you might be inclined to tell us to sprint into oncoming traffic. But before you get too hasty, let us describe a little something that we like to call "duty cycles."

Every part of your body is primed for a certain number of uses, or duty cycles. When you move well, you burn through only one duty cycle with each movement or sustained position. When you move poorly, you burn through *many* duty cycles. Put another way, moving poorly is like overspending from your body's bank account. You have plenty in the bank to make it well into old age (and you can even splurge a lot), but you cannot afford to be deficit spending starting in middle school.

Let's put this into perspective. If you take 10,000 steps a day (the usual recommendation for active people), that's 70,000 steps a week, over a quarter of a million steps a month, and three and half million steps a year. In four years, you'll have taken over 14 million steps. In 10 years, 36 million steps. Now imagine that you've taken every one of those steps with poor technique. You walk like a duck, you roll your foot inward, your arch collapses, your big toe collapses toward your little toe, you wear flip-flops or high-heeled shoes, or you walk with an overextended back—all of which compromise your posture. While the body is designed for a large number of duty cycles, it is not designed to handle an amplified number due to improper positioning. You can think of yourself as an expensive race car. You can still go very fast with the hand brake on, the engine low on oil, and the wheels out of alignment, but eventually you won't go anywhere at all.

We once had a client who came in for physical therapy because she had tweaked her back so badly that she could barely move. She said, "I just leaned over to pick up a pillow off the floor and, bang, my back was tweaked. Weird, huh?" Well, not really. You see, she didn't tweak her back by bending over to pick up that pillow. She had picked up everything in her entire life (her kids, groceries, etc.) in a suboptimal position, heavily taxing her duty cycles. She tweaked her back from years of moving poorly.

Similarly, we routinely see clients in our clinic who are hunched over their cell phones in the waiting room, which is a position we assume they adopt for many hours a day. We think to ourselves, "I bet this person is here for a neck or shoulder tweak." Invariably, they confirm our suspicions. The problem is that no one has made the connection for them before—that adopting certain positions for hours on end every day increases the number of duty cycles used. They can get away with it for a while, until they can't. And then they land in our office. They went one duty cycle too far.

The problem is that we no longer move how our bodies are intended to move, and no one has taught us how to move. We wait until we break down and are in pain before making real change. But by that point, many such injuries are more difficult to fix. Our goal is to teach you how to perform some very basic movements properly so you can avoid catastrophic breakdowns. And, more important, if you do most of your sitting at work or at school, you certainly don't have to let those hours negatively impact the other hours of your life.

In recent years, a segment of the population has caught on to this concept; in many businesses and schools, movement is promoted constantly. While these efforts are certainly beneficial, a key element is still missing: the element of moving *properly*. If you walk with a flexed spine due to years of sitting and then decide to try the latest workout trend without addressing your body position, you are infinitely more prone to movement quality–related injuries. Is moving with improper technique better than not moving at all? Sure. But with every faulty step you take, you're burning through more of your duty cycles. You don't have to choose one or the other. You can have your cake and eat it, too. Our good physical therapist friend Gray Cook has a great saying, "Move well, move often." Notice that he prioritizes moving well first.

To help encourage movement, many forward-thinking companies are starting to offer their employees standing workstations. This is wonderful because standing not only burns more calories than sitting, but also promotes more movement throughout the day. When you stand, you can shift from side to side, alternate between standing on both feet and leaning against a stool, and walk away from your computer for a moment. It's also far easier to take short movement breaks, which add up to a lot of movement over the course of a day. However, few companies instruct their employees on how to stand *correctly*.

Don't get us wrong—watching these transitions happen is wonderful. But to eliminate many of the nasty ailments plaguing modern society, we need to take it to the next level. The problem can be distilled to three simple points:

1. **We are not moving enough.**

2. **We are not moving well.**

3. **We are not performing basic maintenance on our bodies.**

Our society is starting to address the first problem, and this book will offer many insights into how to bring more movement into your life, but the majority of this book focuses on the second and third points. You can teach yourself to move well and have a basic plan for treating your soft tissue restrictions and problems. This is where our system differs. If just moving more were the answer, the millions of elliptical machines on the planet would have turned the tide already.

We are not going to lie—undoing the ravages of a lifetime of working against your body's design is going to require some work. You may have to relearn how to stand, how to sit, and how to walk. You will need to learn how to perform basic maintenance on your body and create new habits. In the beginning, you will need to obsess over your posture by constantly checking and correcting yourself until you have trained your body to move the way it was intended to move. We know this sounds like a lot of effort, but take a moment to envision the upside. Imagine being pain-free and being able to move like you did as a child, and how significantly that will improve your quality of life.

We know that learning about posture, movement, and self-maintenance is about as effective as Ambien for putting people to sleep. It was never our dream to lecture people about the dangers of poor posture or the skills of standing and sitting. In fact, if you were to go back in time and tell our younger selves that our job as adults would be to teach people how to sit and stand with optimal technique, those kids would shed a tear. In our minds, it's much more exciting to teach people how to eliminate debilitating pain or to take world-class athletes and make them better—both of which we do. But realize this: we can't resolve pain or optimize performance without recognizing the negative impacts of sitting and being sedentary and the importance of technique as it relates to position and movement.

To give you an idea of some of the changes you might need to make, we'll offer a brief overview of how this book is organized and how it should be used. But first, let's review four important principles.

The Movement Brain

It turns out that our sedentary lifestyles are not just bad for our bodies, but also bad for our brains. Humans throughout history have realized that physical activity leads to creative thinking, innovation, and optimal cognitive function. The ancient Greeks understood the link between walking and optimizing cognitive function for students. Based on the principle of maintaining a sound mind in a sound body, Aristotle founded the famous Peripatetic School, where teaching was conducted while walking on pathways around the Lyceum.[37]

Dr. John Ratey, Harvard Medical School professor and author of Spark: The Revolutionary New Science of Exercise and the Brain, *writes that the brain responds like muscles. It grows with use and withers with inactivity. "What's even more disturbing, and what virtually no one realizes, is that inactivity is killing our brains—physically shriveling them," he says.*[38]

Many of the world's greatest thinkers—including Winston Churchill, Leonardo da Vinci, Charles Dickens, Ernest Hemingway, Virginia Woolf, Thomas Jefferson, and Benjamin Franklin—were inspired to pen their finest pieces or make earth-shattering decisions while working at standing desks.[39] *George Nelson, iconic lead designer for Herman Miller, liked to stand at work and figured that others probably did, too. He designed a stand-up rolltop desk in the 1950s.*[40]

In one case, British researchers evaluated more than 10,000 subjects between the ages of 35 and 55. Their activity levels were graded as low, medium, or high. The subjects who were more sedentary were more likely to have poor cognitive performance. Notably, those who had low levels of physical activity also had poor fluid intelligence, the capacity to think logically and solve problems in novel situations.[41]

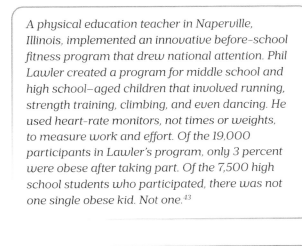

Researchers at the University of Illinois at Urbana-Champaign are conducting cutting-edge brain research on the cognitive benefits of physical activity and aerobic fitness throughout the human life span. Their November 2014 study indicates that the growing emphasis on achieving high scores on standardized tests at the expense of getting a break to exercise may actually be leading to worse academic performance. The researchers found that when compared to their less-fit peers, children who are physically active have larger gray matter brain volumes in the basal ganglia and hippocampus. These brain areas are specifically associated with cognitive control and memory.[42]

A physical education teacher in Naperville, Illinois, implemented an innovative before-school fitness program that drew national attention. Phil Lawler created a program for middle school and high school–aged children that involved running, strength training, climbing, and even dancing. He used heart-rate monitors, not times or weights, to measure work and effort. Of the 19,000 participants in Lawler's program, only 3 percent were obese after taking part. Of the 7,500 high school students who participated, there was not one single obese kid. Not one.[43]

New research contradicts the long-term guidelines for treating children with ADHD and suggests that incorporating standing desks, movement breaks, and more physical activity in general can help these kids perform better. The bottom line is this: all children—especially children with ADHD—must be given every opportunity to move and exercise so that their brains can grow and they can concentrate and learn more and feel better.[44]

Physical activity and movement act like "Miracle-Gro for the brain," according to Dr. Ratey. Exercise and movement don't directly make us smarter or better able to focus. What exercise does is make us more able to learn and focus and optimize the brain for learning. This happens by way of a growth factor called brain-derived neurotrophic factor (BDNF). BDNF allows us to make new connections and learn new material. It spurs neurogenesis—the cell growth necessary to develop the brain in childhood and slow the natural aging process later in life. Even moderate movement (like standing or fidgeting) can supercharge mental circuits and sharpen thinking skills.[45]

Deskbound Guidelines

Preventing and solving the problems associated with too much sitting is, on the surface, very simple. We need to increase our activity, improve the quality of our movement, and learn how to perform basic maintenance on our bodies. That is exactly what this book teaches you to do. The best part is that our system will work for anyone. It doesn't matter if you're hopelessly tied to a chair for 10 hours a day, in chronic pain, or severely overweight. With consistent, conscious effort and a little bit of willpower, you can increase your productivity, lose weight, and treat, avoid, and even eliminate pain. All you have to do is follow four simple guidelines:

1. **Reduce optional sitting in your life.**

2. **For every 30 minutes that you are deskbound, move for at least 2 minutes.**

3. **Prioritize position and mechanics whenever you can.**

4. **Perform 10 to 15 minutes of daily maintenance on your body.**

As you will see, none of these guidelines requires significant change to your current lifestyle. In fact, you will be surprised at just how simple making dramatic improvements to your health can be.

Later sections of this book provide the knowledge and tools you will need to follow these guidelines, but this list clearly spells out your goals for resolving and avoiding the pitfalls of being chained to a desk. In other words, these guidelines won't provide you with all the nitty-gritty details, but they will help you understand why each principle is so important in achieving the ultimate goal—to be healthy and pain-free.

GUIDELINE 1:
Reduce Optional Sitting in Your Life

The title of this guideline pretty much says it all. Sitting only when necessary is one of the best things you can do for your health. When you take on this goal, you will be surprised at just how much of that nasty habit you can remove from your daily routine. We realize that you're probably not going to get rid of your kitchen table, and you still have to drive or ride to work and fly on planes when taking trips, but if you're like most people, you can remove hours of sitting from your day without dramatically changing your lifestyle.

Work is the biggest opportunity for change. Switch to a standing desk if you can. Even if your workspace is a cubicle equipped with a built-in desk, you can throw a couple of boxes on top of the desk, place your monitor on top of the boxes, and work while standing up.

Look for ways to eliminate sitting from your leisure time, too. We certainly aren't suggesting that you stand at the dinner table. But because sitting on the floor is better than sitting on a couch, consider watching TV on the ground or using your couch as a platform and not a chair. When you sit on the floor, you can spend time stretching, squatting, or working on your mobility. It's how we watch TV in our household, and we highly recommend it. It's an easy way to kill two birds with one stone.

When you find yourself forced to sit during the day, do your best to prioritize your body's mechanics. This isn't an all-or-nothing endeavor. Small improvements make massive differences over time. We recognize that some of you might even be wearing a helmet and strapped to a warplane, so do what you can with what you have. We devote a whole section to this topic later in the book—see Section 5. But don't take this the wrong way; just because we think there is a more ideal way to sit doesn't mean that you can forget about eliminating as much sitting as possible. Even when you are sitting in a chair as well as a human being can sit, you're still sedentary.

GUIDELINE 2:
For Every 30 Minutes You Are Deskbound, Move for 2 Minutes

Sitting is bad for a host of reasons, but these two top the list:

1. **Sitting is an orthopedic disaster and can cause a myriad of body dysfunctions.**

2. **Sitting means that you are not moving, and being sedentary can have significant negative long-term health impacts.**

The way to address these issues is to sit less and *move more*. We are fans of standing and standing desks because they create a more movement-rich environment. At a standing workstation, you can easily and constantly change your position and move. If you can't stand at work, then your mission is to move as much as possible–we recommend moving around for at least 2 minutes for every 30 minutes you are deskbound.

The movement doesn't need to be complex. We aren't talking about doing jumping jacks or sprints; we're talking about non-exercise activity, like taking a short stroll around the office, standing and moving your limbs around to promote blood circulation, or, on the extreme end, performing 10 bodyweight squats. We provide an easy blueprint for creating a movement-rich workstation in Section 4, as well as a sample movement and mobility routine that you can utilize to get the most out of these short sessions, but you don't need to read the whole book to get started.

The best part? Movement will likely make you more productive–see "The Movement Brain" on pages 22 and 23. Indeed, Steve Jobs was notorious for holding important business meetings while walking. He knew that the mind does its best thinking when the body is in motion.[46] Business ninja Richard Branson has blogged about his preference for standing or, even better, walking meetings because he finds them to be more productive.[47] Mark Zuckerberg, the founder of Facebook, works at a standing desk and encourages his staff to do the same.[48] These leaders all figured out on their own what research is now confirming: that we think better and are more productive when we are standing or moving than when we are sitting.

We also recommend that you constantly change your position throughout the workday. If you're still at a sitting workstation, we offer some functional positions that you can cycle through on page 196. If you've changed to a standing workstation, we offer various positions on page 145. Standing is the first step, but moving is the goal.

The other great part about adding movement to your day and constantly shifting your position is that it gives you a chance to recapture your posture and reset your body's mechanics. The longer you remain idle, the greater your chance of defaulting to a bad position, such as slouching forward or overextending your back. We'll break down these poor positions later, but let us assure you that making these two simple changes will help you optimize your positioning.

The most difficult part about making these changes is not the movements themselves, but remembering to stop what you're doing and perform them. We get it, you are cranking, in the flow state, on fire. Would you go a whole day without brushing your teeth because you were too busy working? Nope. It's just a change in consciousness. Luckily, there are some wonderful reminder tools out there. The simplest method is to set a timer on your phone that goes off every 30 minutes. There are more complex apps that you can install on your computer, such as ones that literally shut down your display every 30 minutes, preventing you from doing any more work. We list several of the more popular tools in the back of the book (see page 353). It doesn't matter which tool or app you choose, but we highly suggest that you get some sort of reminder in place until moving more becomes a part of your daily routine.

There's no getting around it: if you want to live a long, healthy, and pain-free life, you must move throughout the day. This book will give you some simple and easy tools to accomplish that goal.

GUIDELINE 3:
Optimize Position and Mechanics Whenever You Can

In Section 2, we provide a blueprint for organizing your body into better positions. This blueprint creates a template for moving well all the time, which means that you'll use the same strategies for doing a back squat as you will to bend over and tie your shoes. Learning how to get your body into good positions takes very little effort; most people pick it up in 10 minutes or less. The hard part is making a habit of it. This takes practice. You weren't a good driver the first time you got behind the wheel. It's important to approach organizing your body and performing everyday movements as skills to be developed.

To make better positions your default, you must view moving well as good "nutrition" for your body. We have a lot of clients who acquire this mindset in the gym but then forget to bring this practice into their daily lives. The minute they set foot through the gym door, they constantly check and recheck their posture and are meticulous about how they move. However, as soon as they exit the gym, they leave that mindset behind. We've watched too many clients sit down and immediately slouch over their phones not 10 minutes after practicing immaculate spinal mechanics under challenging conditions. This drives us crazy because the techniques that they perform in the gym are the same movements with the same shapes that they should be performing outside the gym. You squat every time you sit down in a chair. You perform a deadlift every time you bend over to pick something up. If you apply the same good form that you use in your workouts to the actions that you perform in everyday life, including sitting and standing, we are pretty sure that your experience with your mechanical body will improve.

When it comes to mechanics, training in the gym and performing every-day activities are one and the same. Sure, the movements that you perform in the gym are faster and heavier and you may be breathing harder, but the basic organizational principles—the step-by-step models for moving well—are iden-tical. If you differentiate between your athletic self and the person who picks up the kids and gardens in the yard, you're missing the best takeaway from that daily exercise routine, which is how to move efficiently your whole life.

If you're one of those athletes who cares more about making gains in your sport than anything else, let us hit you with this selling point: When we started getting really active about improving our clients' awareness of their positions outside the gym, we noticed huge changes in their athletic performance. Once they eliminated all unnecessary sitting from their lives and started practicing good mechanics at work and at home, their train-ing improved dramatically. Weird, right? Constantly practicing those more functional shapes in their away-from-the-gym lives enabled them to excel in their sports. The instant our more die-hard athletes saw these improve-ments, they began to work even harder on their positioning outside the gym, which allowed them to excel even more. And guess what happened? The bridge separating "training" from "everything else" narrowed. They heard us shout, "Keep your back flat!" every time they deadlifted in the gym. Now, when they go to pick something up at work or at home, they hear our voices echoing those same words. In turn, this improves their deadlifts in the gym. You may have heard the maxim, "The way you do one thing is the way you do everything."

In our book *Becoming a Supple Leopard*, we use the language of strength and conditioning and functional movement—squats, deadlifts, push-ups, and so on—to describe optimal human movement. In this book, we approach it from the other direction. The techniques we will teach you for sitting, getting up and down off the ground, and bending over are the same techniques that you use in the gym. Whether or not you consider yourself an athlete, you need to understand how to squat and pick something up off the ground with-out unduly challenging the structural integrity of your body. You still need full range of motion in your joints and tissues. And you still need to perform basic maintenance on your body.

This book provides you with the tools—the same tools that elite athletes use in their training—to identify and correct inefficiencies in your position, mechanics, and tissue health. When you reach the last page of this book, you will have the same system that we teach to professional sports teams and elite military personnel. You will have all the tools you need to improve the quality of your physical life. Even if you're deskbound, you need to approach your positioning with the same dedication that athletes put into their training. After all, your hips, spine, and shoulders are pretty agnostic about the context in which you use them.

Athletes are motivated by performance, and you should be, too. Be motivated by the idea of moving and feeling better. If that's not your gig, then maybe the fear of an increased risk of cancer or heart disease or of being rendered immobile by age 40 will do the trick. It doesn't matter what motivates you, only that you find a way to get motivated. All roads lead to Rome.

GUIDELINE 4:
Perform 10 to 15 Minutes of Daily Maintenance on Your Body

There are no days off when it comes to performing basic maintenance on your body. "Basic maintenance" refers to the mobility techniques used to resolve pain, improve joint and tissue range of motion, and restore suppleness to tight muscles. On the surface, this looks like classic stretching and foam rolling, but really it's much more than that. Most people know how to get into classic stretching poses, and more and more people are using foam rollers to self-treat tight muscles and alleviate pain. The problem is that people don't have a system for understanding how to perform these techniques correctly. Sure, stretching might improve your flexibility, and rolling a foam roller under a tender muscle might help reduce pain and stiffness. But to maximize your results, you need a set of principles to guide your practice. Section 6 provides those principles.

We use the words *mobility* and *mobilizing* to describe the practice of performing self-maintenance on the body. These terms capture more than just stretching and foam rolling; they encompass movement, suppleness, flexibility, and progress. There is no one-size-fits-all approach when it comes to performing self-maintenance. As with any physical practice, you have to take a systematic approach to attack a particular issue—whether it is an ache or pain, a tight muscle, or an inability to get into a good position. You can follow the guidelines for mobility programming in Section 7 and use them to construct your own personalized mobility prescription, or you can focus on specific areas that need work by following the prescriptions.

Like brushing your teeth and flossing, self-maintenance will be more effective if you commit to a daily practice. The short-term benefit is that you can treat the problem of the day (maybe your back hurts from sitting or your legs are tight from standing all day). The long-term benefit is that you can prevent problems altogether, well before pain and stiffness set in. Rather than waiting for your back to flare up to know that you are sitting too much or doing something else wrong, commit to spending 10 to 15 minutes a day performing basic maintenance.

The key is to be consistent before being heroic. A teacher once told Kelly, "Your muscles and tissues are like obedient dogs. If you spend enough time training, they will respond." If you're struggling to make change, remember that your muscles didn't get that tight overnight. It took years of moving incorrectly and ignoring your range of motion or tissue restrictions for your joints and muscles to become a tangled mess. It will take time, but we've worked with thousands of people, and we've never met one who wasn't able to make significant change with 10 to 15 minutes of daily self-maintenance.

Section 7 includes 14 different mobility prescriptions that you can do throughout your day, because we know that everyone is busy and needs some starter recipes. These prescriptions cover every region of the body.

If you're eager to get started, read Section 6 and then start implementing the techniques and prescriptions in Section 7. But don't skip over the other sections entirely. Bear in mind that if you never correct your positioning, you will continue to experience the same mobility problems. Think about it like this: mobilizing will treat symptoms and might even help you avoid them, but correcting your position will cure the disease.

What Is Mobility?

Many people mistakenly think stretching and mobility work are synonymous. Stretching focuses only on lengthening short and tight muscles. Mobility, on the other hand, is a movement-based integrated full-body approach that addresses all the elements that limit movement and performance, including short and tight muscles, soft tissue restrictions, joint capsule restrictions, motor control problems, joint range of motion dysfunction, and neural dynamic issues. In short, mobilization is a tool for globally addressing movement and performance problems. You will understand this term better after you read Section 6.

How This Book Is Organized

We've broken *Deskbound* into seven sections for easy navigation. Each section offers components and insights that are critical to the next, making it important for you to read the book from cover to cover, at least the first time through. For example, you can't learn how to walk, bend over, or squat (covered in Section 3) or properly set up your standing desk (Section 4) until you learn how to organize and stabilize your spine, hips, and shoulders (Section 2). And you'll never experience the true benefits of mobilizing (Section 7) until you learn the systematic approach for performing basic maintenance on your body (Section 6).

To put it another way, this is not just a book; it's a *system*. And, like all comprehensive systems, the pieces of the puzzle fit together perfectly to form a complete picture, in this case a picture of a healthy, productive, and fit deskbound warrior. Every section includes life-changing techniques and strategies, and we encourage you to revisit each section as often as needed—especially the mobilization techniques and prescriptions in Section 7. Doing so will not only bolster your knowledge, but also help you take your movement and mobility practice to the next level.

SECTION 1:
Consequences of Poor Posture

Compare almost any photo of an indigenous tribesman with that of a modern deskbound human and the biggest difference you'll notice, other than attire, will be posture. When we force our bodies into positions that they were not designed to adopt, day in and day out for years and years, ugly things happen, especially with our spines. The human body was designed to stabilize the spine under normal working conditions, but normal working conditions have changed dramatically since the time of our ancestors, and our bodies haven't yet adapted.

When we compromise the musculature designed to support our spines by assuming positions that are unique to the modern world, the spine searches for its own stability, which often means simply hanging on the end ranges of the body. To make things worse, these adaptation patterns form a sort of "cast" of stiffness around the body that carries over into all aspects of life, not just sitting. This section breaks down the postures that cause the most harm. Once you understand the consequences of poor posture and can easily identify the compromised shapes (flexion and overextension), prioritizing spinal mechanics and correcting the problem (covered in Section 2) becomes a key motivation.

SECTION 2:
Natural Body Principles: How to Organize and Stabilize Your Spine, Hips, and Shoulders

This section offers a simple blueprint for organizing your spine, shoulders, and hips in a stable, anatomically correct position. In other words, it teaches you how to organize your body using a few simple steps so that your muscles and connective tissues are supporting your spine, shoulders, and hips. We also show you how to breathe as you are designed to breathe, through your diaphragm—and how to breathe while keeping your spine braced, or stabilized. This might be the most important section of the book, so don't skip it! Understanding how to organize and stabilize your body in a good position is like unlocking a hidden superpower. You gain access to more power and strength, and nagging pain symptoms tend to disappear.

SECTION 3:
Moving Well: Walking, Hinging, Squatting, and Stable Shoulders

Once you become comfortable organizing and bracing your body in stable positions, the next step is to apply those principles to actual movement. It's wonderful to be able to stand up and organize your body using the strategies provided in Section 2, but if everything collapses the instant you begin moving, then all that work was pretty much for naught. The ultimate goal is to be able to support your spine and work within your body's design in all situations, whether you are walking, hinging, squatting, or carrying something heavy. In a nutshell, this section shows you how to organize your body when performing *baseline movements*—those movements that are at the root of all other movements. For example, walking is a scaled-down version of running, and squatting is a scaled-down version of jumping and landing. As long as you understand how to organize your body for these baseline movements, you can apply that knowledge to other movements that branch off of the baseline movements. Good positions are always open-ended conversations. As skills, they are universally transferable.

When you finish this section, you will understand the ramifications of walking and squatting with your feet turned out, why wearing cushioned shoes compromises your gait, how bending over incorrectly can leave your back open to dysfunction, and how texting with your head down ruins your posture. In short, you will learn the most common faults associated with the baseline movements. We will explain what causes each error—whether it is environmental, technique-based (mechanics), or related to a tissue or joint restriction—and, more important, how to correct the problem.

Some of you will find it difficult to change your poor movement patterns, but we promise that it won't require a dramatic shift in your lifestyle. We are not trying to take you back to the Stone Age. We honestly believe that everyone can reacquire our inherent movements and apply them in modern life with a few simple lifestyle changes and some practice.

SECTION 4:
The Dynamic Workstation

Our goal with this book is to wake you up to the perils of inactivity and give you the knowledge and tools you need to turn your workstation into an active environment so that while you're at work, you're not gaining weight, becoming depressed, causing physical dysfunction, and taking hours, days, or even years off your life. Rather, your work life offers opportunities for getting fitter, stronger, and leaner; increasing your productivity and motivation; and improving your physical performance. To do that, you have to get up from your chair and begin to work while standing up. In Section 4, we'll show you how to set up a standing workstation that is specific to your body's unique dimensions. In addition, we'll outline some strategies for increasing your activity by using simple techniques that you can perform while at your desk.

But you have to take the first step, which is to stand up. The problem is, switching from sitting to standing and moving all day requires some mindfulness and a little progression, which is why we've devoted an entire section to making this transition safely and more comfortably. To help put your mind at ease, we'll outline some tips and tricks to avoid some of the pitfalls that people encounter when transitioning from a sitting workstation to a standing one.

SECTION 5:
Optimizing Your Sitting Mechanics

Our bodies are designed to be on the move constantly—walking, running, squatting, gathering. But in this day and age, it is virtually impossible to remove *all* sitting from our lives. No one is going to walk 30 miles to work, stand up in the back of a movie theater, or take a boat to reach another continent, and by no means do we promote such extreme measures. Many of you have no choice but to sit at work because you are pilots, drivers, police officers, students, or other professionals for whom sitting is not optional. Sitting is here to stay, whether we like it or not. Luckily, not all sitting is created equal.

Section 5 outlines the best ways to lessen the downstream compromises caused by the necessary evil of sitting. First, we describe two less harmful forms: sitting on the ground and passive sitting in a supportive chair. If you have no choice but to sit at work, or if you have to sit for medical reasons, we provide an action plan to help you survive your time in the chair by doing what we call "active sitting." We explain how to apply the organized spine concept to the seated position and why it is important to sit on the edge of your seat. We offer various sitting positions that you can cycle through to inch yourself out of the idle category, as well as guidance on selecting a desk chair. Finally, we give you some tips for enduring car and air travel. Sitting properly is not an easy escape route from the ramifications of sitting, but for those times when you can't avoid long bouts of sitting, this section outlines the best ways to mitigate some of the mechanical effects.

SECTION 6:
Performing Basic Maintenance on Your Body

The mobility part of this book is quite large, so we have broken it up into two sections. Section 6 teaches you the general guidelines for performing regular maintenance on your body to improve range of motion and avoid pain or, if you are already in pain, to address the minor problems associated with our modern sedentary lifestyle. Think of mobilizing as an upgraded version of stretching and foam rolling. The names of these mobilization methods—such as smash, contract and relax, pressure wave, and tack and twist—might sound foreign and slightly intimidating, but trust us, they are just as simple as traditional stretching, only much more effective. To help you get the most out of the prescriptions outlined in Section 7 and to prevent you from causing harm to your body, we also provide five rules to guide your mobility work.

Performing basic maintenance on your body requires a few tools. In order to get started, you may need to purchase a few items. Don't worry, you can get most of what you need for $10 or less, or you can probably make do with stuff that you have lying around the house. In this section, we provide a list of the tools that are specifically designed for mobility, as well as some household alternatives. At a minimum, you should have one roller; one small ball, like a lacrosse ball; two small balls taped together; and one larger ball, like a softball.

SECTION 7:
Mobility Prescriptions

After teaching you the principles for performing daily maintenance on your body in Section 6, in Section 7 we provide the actual mobility techniques that you can utilize to improve range of motion, treat pain, undo joint impingements, reduce muscle soreness, and so on. These techniques are organized into 14 mobility prescriptions, 13 of which are focused on a specific area of the body and the last of which is a catchall prescription for those of us who are deskbound. For example, Prescription 1 is a combination of mobility techniques that focus on the head, neck, and jaw, and Prescription 2 is a combination of techniques that focus on the upper back, traps, and scapulae. These 14 prescriptions enable you to treat or resolve issues with every muscle and joint in your body. Each one takes between 8 and 18 minutes to execute.

Here's a quick breakdown of some of the ways these prescriptions can be utilized:

1. **Take the 14-day challenge:** If you're new to mobility work, we strongly suggest that you perform all 14 prescriptions over the course of two weeks. With a total time investment of only 10 to 15 minutes each day, at the end of 14 days you will have mobilized every region of your body. Not only is this a great way to introduce yourself to all the techniques in the book and give your body an overhaul, but it also will tell you exactly where your restrictions lie. After you complete the 14-day challenge, you will know which areas you need to pay the most attention to going forward. But remember that lifestyle choices and minor injuries are constantly changing the "tight" spots on our bodies, sometimes unbeknownst to us, so it is important to take the 14-day challenge periodically to stay ahead of the curve and resolve issues before they manifest as pain.

2. **Treat a specific condition or restriction:** Taking the 14-day challenge is a great way to locate the trouble spots on your body, but if you are already experiencing a specific condition, such as carpal tunnel syndrome, TMJD, or tension headaches, you will find prescriptions that can help. On page 255, we offer a list of the most common conditions that deskbound workers face, along with the corresponding prescriptions that will help improve each of those conditions. While some conditions, such as TMJD, are limited to a single prescription, other conditions can benefit from multiple prescriptions. You can also attack joint and tissue restrictions that prevent you from getting into good positions. For example, if you take the range of motion tests on pages 224 to 228 and you come up short—meaning that you can't get into a position as demonstrated—you can use the list on page 255 to find the prescriptions that will help you improve your range of motion in that area.

3. **Treat a specific area:** With each prescription targeting a specific area of the body, all you have to do is look at the area map on page 254 to find your mobility prescription for the day.

As we mentioned earlier, if you're eager to get started mobilizing, read Section 6, and then begin the 14-day challenge in Section 7. But again, don't skip over the other sections entirely. As we've said repeatedly, if you don't improve your position and mechanics, you are unlikely to ever get to the root of your body's issues.

A Note for Wheelchair Users

We understand that standing just isn't an option for everyone. There are more than 3.3 million wheelchair users in the United States alone. If you are one of these adaptive workers, know that it's still possible for you to optimize your posture and mechanics from a seated position. While intrinsic support systems may be altered in people who require the use of a wheelchair, the principles behind spinal organization, shoulder mechanics, and body maintenance remain the same. In fact, many of the adaptive athletes who frequent our gym often forget that body mechanics and regular body maintenance are even more important to them given the additional tissue loads induced by their chairs. While every human being is likely to have a unique set of tissue restrictions, everyone is subject to the set of compromises imposed by chair-shortened and stiffened hips, for example. If you are a wheelchair user, this book is still very much for you and your family.

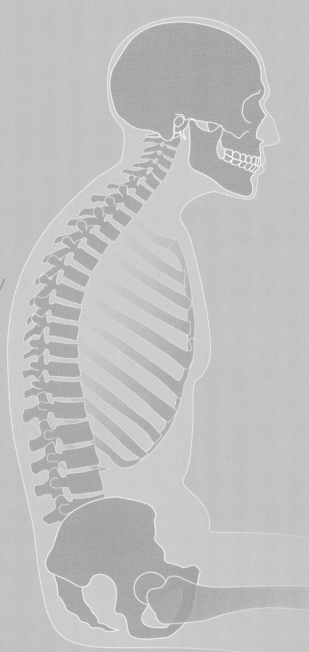

Consequences
of Poor Posture

The human body is incredibly robust. This is a good thing if you have a plan for getting organized. It preserves joints and tissues for a lifetime of use. The problems occur when you don't have a strategy for stabilizing your body in a good position. Your body's design has redundancies and backup emergency protocols built in to keep you moving through your environment. Don't have a plan for how to organize your spine before you lift that piano? No problem; you can lift it any way you want . . . until one day you can't. Don't have a plan for standing at your computer for eight hours straight? It's all good; you can stand however you want . . . until your lower back glows with pain.

What you need to understand is that your spine is always searching for stability. Imagine holding one of those wooden snake toys by the tail and trying to balance it vertically. You are going to work really hard until you give up and let the snake round to the limits of its wobbly wooden sections. This is what happens to your spine when you fail to organize and stabilize your body in a good position: the bones that make up your spinal column (your vertebrae) act like the wobbly wooden sections of the toy snake. So, if you fail to provide stability for your spine by actively using your musculature and tissue systems, your body will attempt to create a sort of second-tier, or reactionary, stability on its own. In other words, if you don't stabilize your body the right way, it will default to lesser-quality stability by rounding forward (flexion) or arching back (overextension). This might be useful if you suddenly have to move without warning, but flexion and overextension, the two primary spinal faults covered in this section, are inefficient forms of stability that wreak havoc on your body.

Now, if you're like us, you probably care less about what *not* to do and more about what you *should* do. And we promise that we'll get to the "what you should do" part in the pages to come. But first, we want to make sure that you understand why the postures associated with sedentary behavior are less than ideal. Why is this important? Because once you realize the consequences of poor posture and can spot it in yourself and in others, the not-very-exciting topic of spinal mechanics becomes an instant priority.

In this sense, understanding some of the downstream consequences of poor posture can help you see why this is an important—and easy—problem to solve. Moreover, knowing the potential causes and effects can help you grasp the relationship between some of the common aches and pains that we experience and the shapes that we adopt.

The Rounded Spine:
A Flexion Fault

7 cervical
vertebrae

12 thoracic
vertebrae

5 lumbar
vertebrae

← sacrum

← coccyx

Walk into any office building and you will likely see the majority of workers sitting at their desks with their backs rounded forward, their shoulders caved in, and their heads hanging out in front of their bodies . . . all of them looking as though they are suffering from advanced stages of osteoporosis, depression, and old age. The beauty of your S-shaped spine made up of interconnecting vertebrae is that it gives you the gift of movement—the ability to bend, arch, and twist. But your spine can also be your biggest weakness. Slouching forward in your chair and forcing your spine into a C-shape for prolonged periods compromises your spine's stability, integrity, and innate function. Yes, you *can* eat chocolate donuts with every meal and chase them with a pack of cigarettes, but you know intuitively that this isn't a solid long-term nutritional plan. The problem is that we don't apply the same level of scrutiny to our body's shapes.

To help you understand what we mean, let us take you on a walk in the woods. Look around you at all the pine trees. See how their trunks are almost perfectly straight? Although each tree has thousands of pounds of limbs and needles, all that weight is supported by that straight trunk, which in turn is supported and stabilized by an expansive root system. Those trees can withstand incredible winds and heavy blankets of snow because they are braced. However, every once in a while you will find a tree reaching out over a river in an attempt to grab more sunlight. Like the straight-trunked trees, this tree also has thousands of pounds of limbs and needles, but what is supporting it? With the trunk extending horizontally, the bracing element is gone. The trunk must withstand incredible pressure from all that weight, as must the root system clinging to the shore. Add snow and wind to the equation, and all that unsupported weight often is too much for the tree to bear. Either the trunk snaps in half or the roots are ripped from the ground.

When you sit down, the musculature of your lower body basically turns off. This is one reason your ankles swell on a long airplane ride. With your lower leg muscles no longer squeezing and pumping the lymphatic fluid in your leg tissues, you start to get backup or congestion. Try this simple experiment: while sitting down, squeeze your glutes. It's really difficult, right? And even if you can do it well, it certainly isn't very sustainable. When you are sitting, your gluteal muscles are essentially asleep.

This is a problem because your glutes are your root system. With those giant muscles out of play, it's nearly impossible to manage the relationship between your pelvis and your legs. And with this keystone of stability knocked out, you are left using just your trunk musculature to keep your spine supported in its natural shape. The problem is that using just your trunk to support all that top-heavy weight is exhausting, so eventually you relax your abs and back and the whole system collapses.

Don't believe us? Get into a meditation pose and let us know when your back starts aching. Yep, that's right when you would begin to slouch. And this is our point. Traditional desk-and-chair architecture sets us up to fail from the start. Kelly was once lecturing at the Stanford University School of Medicine and challenged the young doctors in the audience to maintain a solid spinal shape throughout the lecture. He only had to remind them four or five times to stop rounding forward.

Even with little support from the surrounding musculature, your body will try to find positions of stability when you slouch. Unfortunately, without those active systems, it must resort to another, less sustainable form of stability, something one of our senior war-fighter friends in the U.S. Navy refers to as "hanging on the meat." Not a very pleasant term, to be sure, but it's an accurate description. With your muscles lengthened and essentially turned off, you're passively relying on soft tissues like fascia, ligaments, and tendons for your spine's primary stability. It's basically a train wreck inside you: your rib cage collapses, your shoulders roll forward, and your head extends away from your body. Hanging on your meat does provide support, just not the type of support that promotes peak function, spinal longevity, and a pain-free life. Essentially, you become that pine tree reaching out over the river. Instead of your top-heavy weight being absorbed evenly by your spine, that weight is concentrated on a certain segment of vertebrae—your cervical spine (see illustration, opposite).

What are some of the physical implications of slouching forward in a position of flexion, you might ask? If you're deskbound full-time and you've been sitting for most of your life (which most of us have), the list of potential negative consequences is long and definitely not limited to the spine. Let's explore some of the most common problems and find out exactly what this typical sitting posture is doing to our bodies.

This tree is not braced. A heavy snowfall could easily cause it to either splinter in half or simply fall into the water.

Despite having very narrow trunks, these trees are perfectly braced and will be able to withstand the elements.

The Rounded Spine

Slouching, hunching, rounding forward, rounded spine, camel hump, kyphosis—these are all terms that are used to describe the flexion spinal fault.

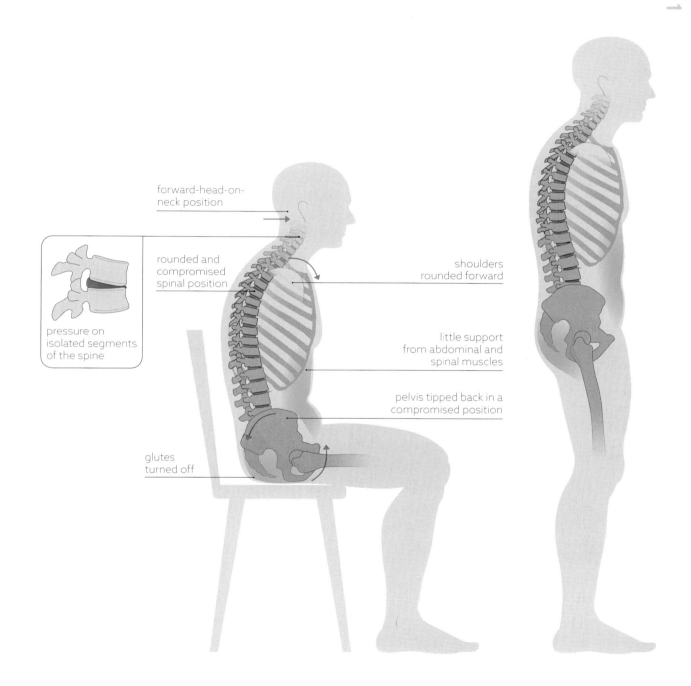

forward-head-on-
neck position

rounded and
compromised
spinal position

pressure on
isolated segments
of the spine

glutes
turned off

shoulders
rounded forward

little support
from abdominal and
spinal muscles

pelvis tipped back in a
compromised position

Loss of Normal Range of Motion

If you constantly sit, stand, or move in a rounded-forward position of flexion, your shoulders, chest, and neck muscles will likely become adaptively stiff, which means that those muscles will tighten into a shortened position. Over time, this cultivated stiffness can compromise your ability to organize and stabilize your spine. It's a slippery slope. The further you slide, the more range of motion you could lose. It's as though your body forms a cast around that flexed position. With your muscles constantly pulling your shoulders and head forward into that unnatural C-shape, it gets harder and harder to reclaim a good position.

If left unchecked, a flexed spine can eventually lead to postural kyphosis—an exaggerated rounding of the upper back, like a camel's hump. The good news is that your body is quite remarkable at remodeling itself back to its normal shapes when given a little input from you.

Quick Loss of Range of Motion Treatment

If rotating your shoulders back into a good position seems like a lot of work, chances are your chest and shoulders have adapted to your slouched position. To restore normal function to these adaptively stiff chest and shoulder muscles, try implementing Prescription 3 on page 266. Then follow the bracing sequence on pages 82 and 83 to correct your position.

Kyphotic Spine

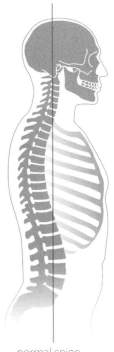

normal spine

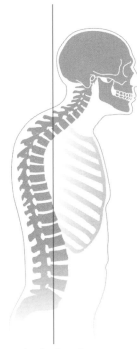

kyphotic spine

Diaphragm Dysfunction

In addition to compromising your ability to move, adopting a C-shaped posture wreaks havoc on your breathing mechanics. That's right: a rounded back can compromise your ability to breathe fully and efficiently. The moment you collapse forward, your diaphragm—a sheet of skeletal muscle that plays a critical role in respiration—becomes compressed and "positionally inhibited." Put simply, when you're in a flexed position, your diaphragm can't function like it's supposed to. You are mechanically compromised by your shape. Unable to take big, rhythmic breaths through your diaphragm, you compensate by taking short, shallow breaths through your neck and chest. This tricks your body into thinking that you are in a state of fight-or-flight, which triggers the release of stress hormones and compromises your ability to down-regulate, or get into a relaxed state. (No wonder you're having such a hard time getting to sleep at night!) And poor breathing patterns are only the tip of the iceberg. Compromised diaphragm function can also result in exacerbation of breathing disorders like asthma and chronic obstructive pulmonary disease (COPD).

In Section 2, we'll talk more about the negative consequences of compromised breathing, as well as provide you with a blueprint for breathing correctly. But let's not get ahead of ourselves. Compromised breathing mechanics is just one of many problems linked to poor posture.

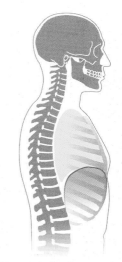

Normal diaphragm

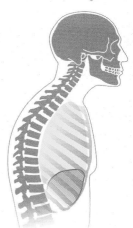

Collapsed diaphragm

Numbness and Tingling

Do you ever get numbness and tingling in your arms and hands? Most desk-bound keyboard warriors do. The moment you round your upper back, your shoulders internally rotate into a forward, collapsed position. But your head remains level, resulting in a visible crease or hinge at the base of your neck. This creates a kink in the tube that houses your central nervous system (your spinal cord), which controls the activities of your body. Your rounded upper back increases the overall tension in your nervous tissues. With your compromised, shallow neck breathing (described above), the muscles of your neck produce even more tension on the nerves running from your neck to your arms. Resting your wrists on your desk and bending them out of neutral alignment—the position in which your joints are naturally aligned—places yet another stress on the global system. Eventually, your body begins to protest these sustained shapes by sending you important signals that your nervous tissues aren't getting enough circulation, in the form of the familiar tingles.

**Numbness and
Tingling Fix**

If you have numbness and tingling in your arms or hands, you can treat the symptoms by performing Prescription 3 on page 266, Prescription 6 on page 284, and Prescription 7 on page 290.

Neck Pain and Headaches

Neck Pain?

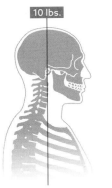

Normal posture

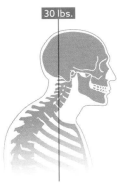

2 inches forward ⟶

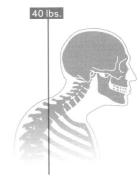

3 inches forward ⟶

Another by-product of a poorly organized, rounded spine is that your head falls out in front of your body. We call this head dangle the "dreaded forward-head-on-neck position." It creates a major hinge in your cervical spine, right at the base of your neck. Instead of positioning your head over your shoulders, which evenly distributes its weight over your entire spinal column, you create a shear force—a perpendicular or off-axis force—across just a few motion segments of your spine. If you hinge from one or two vertebral segments, the weight of your body compresses that hinge. Given enough time in this compromised position, you're bound to spend some time in the compensation-pain cave.

How much neck shear does the forward-head-on-neck position cause? Or, in more relatable terms, how much of an impact does it have? Well, the average human head weighs about 10 pounds. For every inch that your head moves out in front of your body, you add an additional 10 pounds of pressure on your spine.[1] So, if your head is positioned 3 inches in front of your body as you hunch over your keyboard or phone, you are adding 30 pounds of pressure to those segments of your spine and the supporting tissues. No wonder your neck radiates with pain and your upper back and shoulders are tight! Though your head doesn't weigh 40 pounds, your poor position is exerting 30 extra pounds of pressure on your cervical vertebrae. And only 3 inches of forward translation of the head is generous. Most of us are texting or typing while wearing the equivalent of a six-year-old child on our necks.

Your neck muscles are also going to take a beating in this forward-head shape. Their job is to support your head by creating a contraction force, but if they are elongated and forced to work at end range—the limit of your joint and tissue range of motion—for many hours, it is much more difficult for them to do their job. Now they not only have to support your head, but also must do so from an elongated position. Asking muscles to do a job that they were never intended to do creates serious resting tension. Ever have a stiff neck after working diligently on a deadline? We are talking to you. It can also lead to headaches and leaves you susceptible to an upper trapezius tweak (you know, that knot in your upper back that radiates pain up your neck and makes it seem impossible to look left or right). In a nutshell, flexion can cause shoulder and neck problems galore.

Quick Neck Pain and Headaches Treatment
To prevent or treat a tension headache or neck pain, or if your upper back and neck are stiff due to the forward-head-on-neck fault, try implementing Prescription 1 on page 256 and Prescription 2 on page 260.

Low Back Pain

While sitting in a C-shaped position of flexion does a number on your upper back, neck, and shoulders, your lower back also pays a hefty price. With your glutes turned off and your spinal support system shut down, your lumbar spine collapses and creates a lopsided compression force across the discs of your lower back. In simpler terms, your discs get pushed backward (posteriorly), which can cause your vertebral discs to sag or bulge and become deformed over time. And because most of us spend so much time in rounded postures, our discs don't get the chance to rebound; they remain unevenly compressed.

Sitting with your pelvis tucked underneath your body—a position known as "posterior pelvic tilt"—only compounds these problems. Essentially, you're sitting on the bottom of your spine rather than on your ischial tuberosities (the bony bottom of your pelvis that is designed to be weight-bearing). Posterior pelvic tilt combined with the dramatic C-shaped curvature of your spine can deliver a nasty one-two punch to the structural components of your lower back. The curve of your upper body pulls your lumbar spine forward, while your pelvic tilt pulls your lumbar spine backward. The discs themselves may not be immediate pain generators, but don't forget about all the out-of-position muscles and tissues supporting your spine. These abused workhorses have the ability to complain bitterly.

Posterior pelvic tilt is common when sitting on a cushioned surface, such as a couch, padded office chair, or car seat. In Section 5 we describe how sitting on a hard surface can help eliminate this issue. To get yourself heading in the right direction, imagine that you have a tail growing out of the bottom of your spine. Every time you sit down, think about not sitting on your tail.

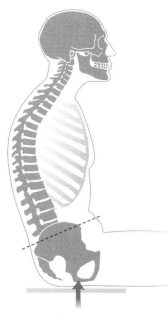

Posterior Pelvic Tilt

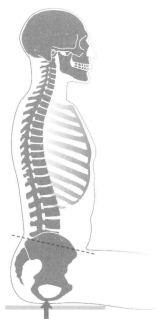

Quick Low Back Treatment

Whether you're imprisoned in a chair for extended periods or your low back hurts, performing Prescription 5 (page 278) will help treat pain symptoms and keep your trunk supple and healthy. Prescriptions 8, 9, and 10 on pages 298 to 319 are also great sequences to employ when suffering from low back pain.

Jaw Pain (TMJD)

Unfortunately, the orthopedic problems that stem from hunching are not limited to the neck and back. A poorly aligned head means that the tissues that run and support your jaw can't work the way they're supposed to. Think of it this way: many of the muscles that operate your jaw arise from your neck. If your head is positioned forward of your neck, your lower jaw gets pulled backward, creating a tug-of-war with the structures of your mouth upstream.

To test this idea, assume a forward-head-on-neck position by slouching and looking down while keeping your jaw closed. Pay close attention to what happens with the tension in your jaw. As your head drops forward, your mouth wants to open. Now imagine spending several hours a day in this position, breathing, talking, and chewing. The muscles of your neck, jaw, and face have to work a lot harder to support your position and as a result get tight and overworked. And your jaw joint being pulled into strange positions can lead to a painful condition known as temporomandibular joint disorder, or TMJD.

As if that weren't enough, a poorly aligned airway often changes your breathing pattern to mouth breathing. Breathing through your mouth tends to reduce the total volume of air that you exchange. This means shallow, inefficient breathing mechanics.

Quick TMJ Mechanics Treatment

Prescription 1 on page 256 is the best sequence for treating and resolving TMJD. It's a catchall prescription for head, face, and jaw health.

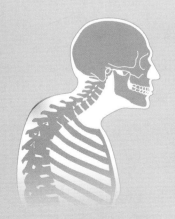

Fatty Neck Hump

While several things can cause a buildup of excess fat behind the neck, one of the big ones is sitting in the forward-head-on-neck position for prolonged periods. In an attempt to even out the forward weight distribution of your head, your body will create a fat deposit under the skin of your upper back. This fat accumulation acts like a tent pole for the tightly stretched connective tissues on the back of your neck. Your body is trying to pull your head back into position by increasing the volume of fat tissue under your skin. And it's not very cute.

The Arched Spine:
The Too-Much-Extension Fault

"Sit up straight." If you're like us, you probably heard this a thousand times growing up. Everyone, from your parents to your teachers, wanted you to sit up and stand up straight.

For a long time people have intrinsically understood that slouching is bad for us, but not necessarily for the reasons covered in this book. Historically, slouching represented laziness, weakness, and insecurity. Your elders wanted you to sit up or stand up straight because good posture exudes confidence and alertness. Go back to the turn of the 20th century and you'll find that even the cutting-edge Art Nouveau chair design of Charles Rennie Mackintosh had a straight-up-and-down back.

For reasons already discussed, sitting up straight is a good thing, but there can be too much of a good thing. While "sit up straight" is an incredible meme about spinal health, it turns out that *how* you sit up straight matters.

Standing and sitting up straight *should* mean establishing a well-organized and physiologically sensible position for your spine (generally referred to as "neutral"), but in most cases it is interpreted as "neutral plus a little." It's the "plus a little" that poses a problem. Cue a child or an adult in a flexed position to stand up straight, and she will most likely rotate her pelvis forward (known as anterior pelvic tilt) and puff her chest out, flipping her C-shape to the opposite side. Most of the time, people don't even need a cue; it happens naturally.

Here's how it plays out: You're working at your computer, focusing on the task at hand, not really paying attention to your position. After a few minutes, you start to slouch in your chair or, if you're standing, you hunch forward over your desk. Once you become aware of your lost good posture—either because you're uncomfortable and your body begins sending you signals to move or because you want to correct your positioning—you straighten up and try to get your back flat. If you had a plan for getting organized, you would reset your pelvis into a neutral position by squeezing your glutes. But most of us were never taught to have a plan, so we try to find stability by overextending our lumbar spines.

Most of us think that the spine ends at the top of the hips. While it certainly is true that the vertebrae end here, the functional spine actually includes the pelvis, just as it includes the head at the other end. When you sit or stand in a way that doesn't utilize your legs very well, your pelvis becomes just one more big vertebra, or motion segment in the spinal chain. In the case of overextension, the vertebrae in your lumbar spine have hard, bony stops, pinching the facet joints that interface between adjacent vertebrae. This sustained compression of the lumbar facet joints can cause a ton of problems, including low back pain and tight muscles.

Because your spine is a miraculous and tolerant mechanical system, it's easy to confuse the very useful and vital bony extension-limiting structures of your spine with the key mechanisms of spinal support. It certainly feels stable when you overextend and lock your lumbar spine into end range. And that wonderful bony block means that you don't really have to utilize any of the active structures that support your nervous tissue column. But we already have a pretty acceptable meme around hanging on the end ranges of our joints. Have you ever heard someone tell a child not to lock (hyperextend) her knees when standing? It's the same concept. The bony end-range protection that we get from our joints keeps us from literally snapping in half, a

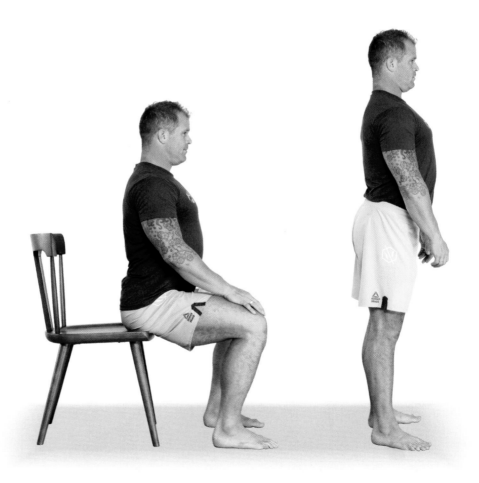

useful evolutionary adaptation to be sure. Utilizing this bone-on-bone strategy in everyday life, however, is like using your bumpers to stop your car. You can do it, but it's not a sustainable solution. And keep in mind our "practice makes permanent" model. If you spend many hours a day practicing an overextended, pelvis-dumped-forward position, how do you think you might move while carrying out the everyday activities that we perform as human beings?

Think of your pelvis as a bowl of water. You don't want to spill any water out the back (in the case of flexion) or the front (overextension) of the bowl.

The Arched Spine

Overextended, swayback, arched spine—these are all terms that are used to describe the too-much-extension spinal fault.

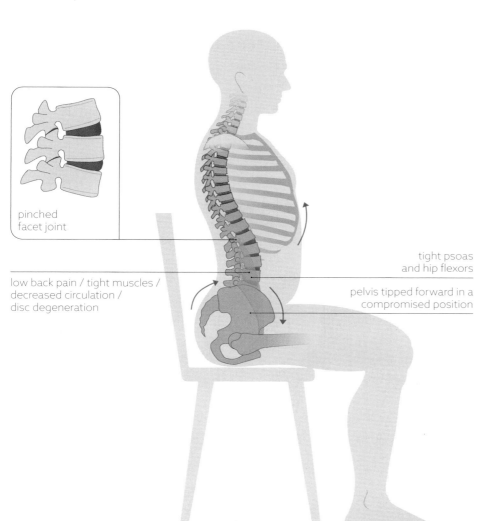

pinched
facet joint

low back pain / tight muscles /
decreased circulation /
disc degeneration

tight psoas
and hip flexors

pelvis tipped forward in a
compromised position

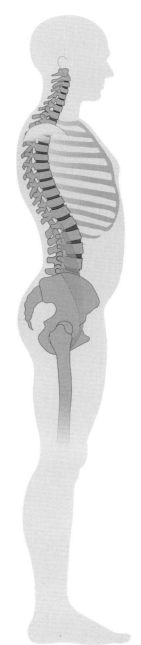

Low Back Pain

The moment you tilt your pelvis forward in an attempt to stand or sit up straight, you create a pretty strong shear force (see page 46) between your spine and your hips. Instead of your upper body being supported by your entire spinal column and all the musculature of your trunk, it is supported by a few jammed-together, localized motion segments of your spine. In our athletic culture, even coaches are guilty of cultivating this dysfunctional pattern, largely because it's easier to tell people to park their lower backs in a cranked-over position of extension than it is to teach them how to brace and stabilize effectively. Additionally, the bone-on-bone "swayback" shape isn't rounded, so the issue of flexion-related disc injury is superficially solved. There are a whole host of problems with this thinking, but fundamentally we have to remember that when the body is functioning at a high level, its basic geometry has to be organized or significant losses of function and output will occur.

Let us give you an example. Our friend John was visiting our gym during the NFL off-season. He is a very lean 6-foot, 5-inch, 310-pound offensive lineman who was in his tenth season in the league at the time. John and Kelly were performing some heavy strongman yoke walks, which involve carrying many hundreds of pounds on the shoulders using a contraption that looks like an upside-down letter U. John is a monster, and one of the best athletes we've ever met, but with a load of only about 600 pounds, Kelly was able to out-walk him for distance every time. Why? John was over-extended, and his less stacked and organized lumbar-pelvis relationship caused him to lose force before Kelly did. Don't get us wrong: John is much stronger than Kelly, but Kelly was simply more effective in transferring that load through his organized spine.

When we treat someone in our physical therapy practice for low back pain, or when we see an athlete who is constantly tweaking his back during training, the first thing we address is spinal mechanics. We view the spine as a carriage or chassis for the big engines of the body. When someone has a poorly organized spine, we see poor upstream and downstream control of the shoulders and hips. Good movement always begins with the spine. We even define functional movement as starting at the spine (proximally) and then moving out to the arms and legs (distally).

In most of the people we work with, the opposite is true: they initiate movement in their extremities. This is like the tail wagging the dog. The problem with a poorly organized spine is that it forces violations of the core-to-extremities rule. For example, girls and women who are comfortable adopting a forward-tilted pelvis exhibit far poorer knee control when they jump and land. What looks like a weak athlete is really just an athlete moving with disorganized architecture, like our footballer friend John.

The downstream issue of using your spine's emergency backstops as a movement strategy is that it places constant and undue stress on tissues that were never intended to be used in this manner. Once you've burned through your lumbar spine's duty cycles, you can become what is called "extension sensitive." In that case, extending too far can result in pain and dysfunction. It is normal to lose some disc height as you age. But if you live and move with a spine that is biased toward extension all the time, coupled with the normal loss of disc height, the bony canals through which the nerves of the spine exit often become smaller and more obstructed. The medical term for this is *stenosis*.

Suffice it to say that stenosis should be thought of as a continuum. Closing a healthy back into an overextended position makes those bony tunnels narrower. Thank goodness those bones keep you from completely pinching off the nerves! Extension sensitivity issues begin to rear their ugly heads when you start to see changes in bone health in these areas, as well as "laxity" in the ligaments that you've been relying on. If your beautiful and dynamic nervous system tissues (nerves) have to start sliding through tunnels that were once smooth and wide and are now rough and narrow, you may begin to see the ramifications of that seemingly innocuous overextended spinal stabilization strategy.

Quick Low Back Treatment
Wondering what you can do to treat or, even better, resolve back pain? Perform regular maintenance—see Prescription 5 on page 278. Prescriptions 8, 9, and 10 (pages 298 to 319) are also effective for treating low back pain symptoms.

"Tight" Muscles

In addition to negatively affecting your spinal health, sustained overextension can cause your muscles to compensate in disadvantageous ways. For example, most of your abdominal musculature becomes positionally inhibited, or rendered ineffective due to your poor position. Since the spine is so important to our survival, we have many redundant systems. When we lose our primary methods of stability, we can use our backup systems, like the larger "primary movers" or other large muscles. With your glutes and abs out of the picture, these bigger engines (muscles) are required to perform double duty: they have to stabilize *and* move. They show their discontent through stiffness, tension, and potentially pain.

Why not give it a test? All you have to do is sit or stand with your pelvis tilted forward. You'll probably feel the big muscles of your back activate. In this same belly-out, pelvis-tilted-forward position, try to engage your abs. You'll probably find that you have to move out of that position before you start to feel your abdominal musculature turn back on in a meaningful way. If you've ever wondered why you have such a stiff back after standing or sitting for a long time, the answer might be your overextended posture.

Your psoas is another good example. This muscle is responsible for flexing your trunk to your hips, moving your hips to your spine, twisting your upper body, and stabilizing your spine. The moment you overextend, your psoas loses its mechanical advantage on your organized spine and begins to exaggerate your normal lumbar curve. The psoas literally drags your lumbar spine's curve toward your pelvis.

To put this in perspective, imagine holding a two-pound weight in your hand with your elbow bent at a 45-degree angle. You may not feel any resistance for the first minute or two, but hold that weight for five minutes and you will start to feel the burn in your elbow and biceps. Now hold that position for 10 to 12 hours a day. This is analogous to what is happening with your psoas when you sit in an overextended position. The weight of your unsupported torso creates resistance that your psoas has to work hard to support. This is what we refer to as a "passive load." Spend enough time in this position and your psoas will do what it has to do to support your poorly organized and stabilized spine. It will get stiff. You've experienced this stiffness when standing up after a long bout of sitting. Initially it's difficult to stand all the way upright. To add insult to injury, that stiff psoas will now be constantly pulling the segments of your spine toward extension, even if you are being conscious of your spinal organization. That means your body's musculature is playing tug-of-war on your spine. It's like driving with your emergency brake on. Wonder why the transition to a standing desk (see Section 4) is so difficult at first? This is one of the reasons.

Tight Psoas Fix

If your psoas is tight or your trunk is stiff from sitting or being in an overextended position, performing the techniques outlined in Prescription 5 on page 278 will help get you back to baseline.

"Tight" and Stiff Muscles

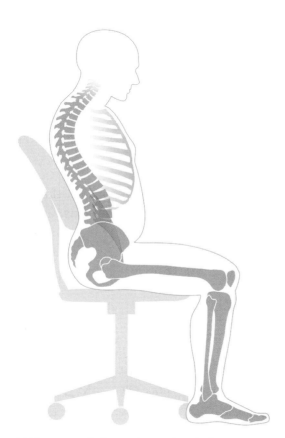

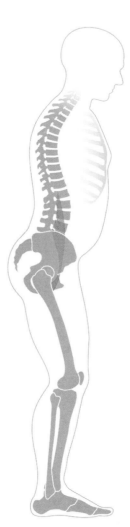

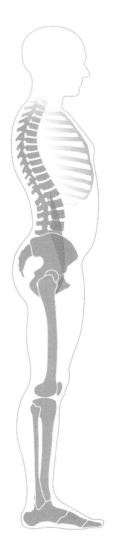

1. When you sit for prolonged periods, your psoas muscle becomes adaptively stiff.

2. And when your psoas gets stiff, it pulls your torso forward.

3. Now, when you stand upright, you can't fully extend your hips, locking you in an overextended position.

THE SIDE SLOUCH:
The High Hip /
High Shoulder Fault

If you were to visit an office building and observe the deskbound workers sitting with typical posture, you would likely see them switching back and forth between flexion and overextension in an attempt to find relief. When both positions become intolerable, you might see them assume a third, even less effective position, which we call the "side slouch." From a position of either flexion or overextension, you shift your weight to one side or the other by tilting your head, elevating one hip, or dropping one shoulder. Imagine a precocious child throwing a fit and trying to emulate a fashion model at the same time.

The side slouch posture shares all the same potential consequences as sustained flexion and overextension: low back pain, compromised breathing mechanics, loss of mobility, muscle imbalances, and tissue restrictions, to name a few. If you find yourself putting more weight on ("biasing toward") one hip when sitting in a chair or posting on one leg and hip when standing, then expect side-to-side imbalances and restrictions as well.

People regularly ask us why just one of their feet turns out like a duck's when they walk or why they seem to bias toward one side when they squat. The answer is that they have programmed those tissue asymmetries through hours of repetition. There is good research to show that an athlete who expresses asymmetric movement patterns is almost three times more likely to sustain an injury. Side standing or sitting is really just another way of getting around the spinal organization problem, but it can greatly contribute to secondary dysfunction when those tissue compensations are challenged during daily life. When we coach young athletes, the first thing we teach them when upright is to stand evenly on both feet with their toes pointed forward.

The key is to be able to identify the side slouch for what it is: a body hunting for a stable, sustainable shape.

Side Slouch

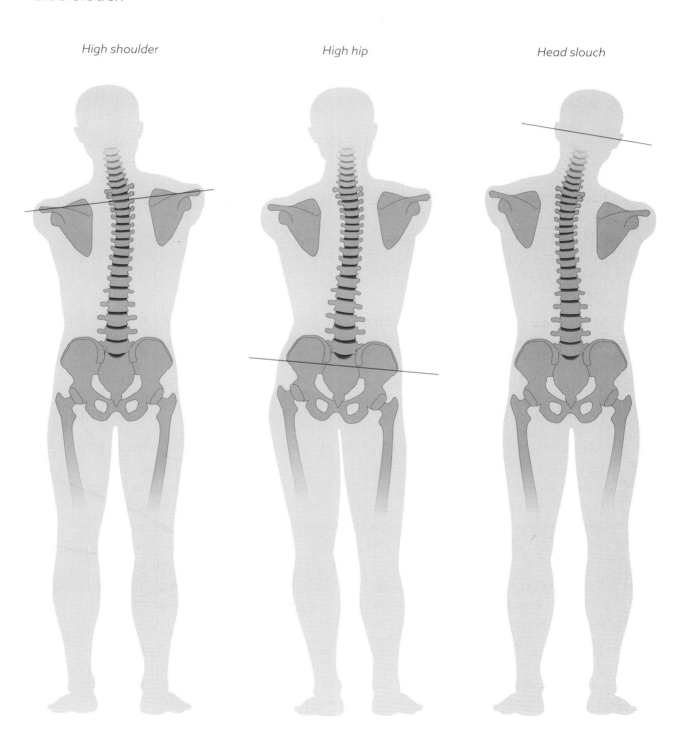

High shoulder

High hip

Head slouch

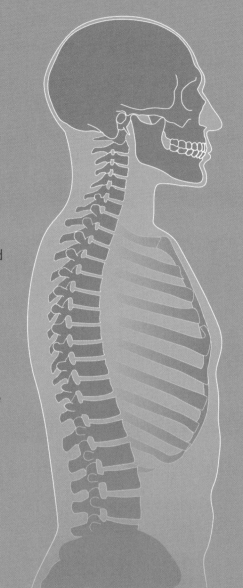

Natural Body Principles:
How to Organize and
Stabilize Your Spine,
Hips, and Shoulders

The human body is designed to readily adopt and maintain an upright position. But our modern habits kill some of that natural capacity and skill. For example, when you sit slouched forward, it is difficult to break that habit when you stand up. And when you sit for prolonged periods, your hip flexors, hip capsules, and trunk muscles become adaptively stiff, which makes it even harder to stand correctly.

When standing correctly proves difficult, *all* movement becomes more difficult. After a while even easy, everyday movement puts strain on your body and hurts your spine, hips, and legs. So what do most people do? They attempt to relieve the pressure by sitting down again, which only exacerbates the problem.

It's a shame that standing has become a skill that needs to be taught. But because standing is the gateway to movement, it makes sense to get it right. Once you learn how to stand properly, you can apply this most basic of skills to most of the movements in the human vocabulary.

We fully understand that for many people, reacquiring the skill of standing might not seem very sexy. And it might not happen overnight. Later in this book we offer a host of mobility techniques that will help you undo the ravages of sitting and reclaim your natural biomechanics, but the most important thing at this stage of the game is to understand exactly what it takes to stand properly. In other words, you need to know how to organize your entire standing body in a safe and sustainable way.

Your body is designed to last a hundred pain-free years, but you can't begin to harness the full range of human function if you don't have the basic principles down first. You need to learn to walk before you learn to run, and that begins with learning how to organize and stabilize your spine.

It doesn't matter whether you are a member of an elite military unit or a middle-school volleyball team, whether you are heading to the Olympics or just going for a walk; it all begins with the spine. And this is where we always start the conversation. The bottom line is that we can't resolve anyone's pain, joint and tissue restrictions, or dysfunctional movement patterns until we address how to stabilize the spine in a neutral position. If we, as a culture, are going to move beyond the "quick fix," then we have to start with the basics.

The Importance of an Organized and Stable Spine

Let's return to our walk in the woods for a moment. (Flip back to page 40 for a refresher.) Surrounding you are towering pine trees with straight trunks and elaborate root systems plunging into the earth. In winter, thousands of pounds of snow cling to the branches and needles of each tree. Despite a harsh wind whistling through the forest, the trees stand tall and strong, just as they have for decades.

Now think of your body as a pine tree. Instead of branches, you have arms, and instead of a crown, you have a large head holding a heavy brain. With all that weight up top, your trunk (that is, your spine) needs to be organized and stable. Although you don't have a root system, you do have your own bracing tools: the musculature of your trunk and lower extremities. These muscles, combined with the incredible connective tissues of your body, create a "root system" that is capable of incredible feats of artistry and output.

Instinctively, we all have the know-how to brace our spines to support a lifetime of use. For example, we rarely see a young child running or playing with a slouched-forward posture. Just like pine trees, we have all the tools we need to thrive in our natural environment. And this "environment" point is the key. The human body is easy to develop and nurture in an environment that constantly requires us to express the full range of motion of our joints and tissues. For example, in Japanese nursing homes, where residents sleep on the ground (and have to get up and down from the floor every day), very few falls are reported. In fact, the number-one reason people end up in nursing homes in the United States is that they can no longer get up off the ground independently.[1] This isn't an issue of ignorance or laziness; it's a matter of our environment not encouraging us to move.

When we keep our spines organized and stable, our bodies are capable of amazing things—like lifting heavy weights. And we can accomplish such feats our whole lives, with few serious consequences. Unfortunately, poor posture and prolonged sitting can override these natural instincts and cause a great deal of dysfunction. Breaking those habits and mitigating the consequences can be difficult, but in this section we lay out a clear and concise blueprint for standing the way our bodies are intended to stand.

The Braced Neutral Spine

This illustration offers a quick breakdown of what bracing your spine in a neutral or natural shape looks like and entails. Don't worry if it doesn't make complete sense right away. Later in this section we will give you step-by-step instructions for bracing your spine in a neutral position, and in Section 3 we will teach you how to maintain that position while moving. But for now, it is important to understand how the spine evolved to be organized and stabilized.

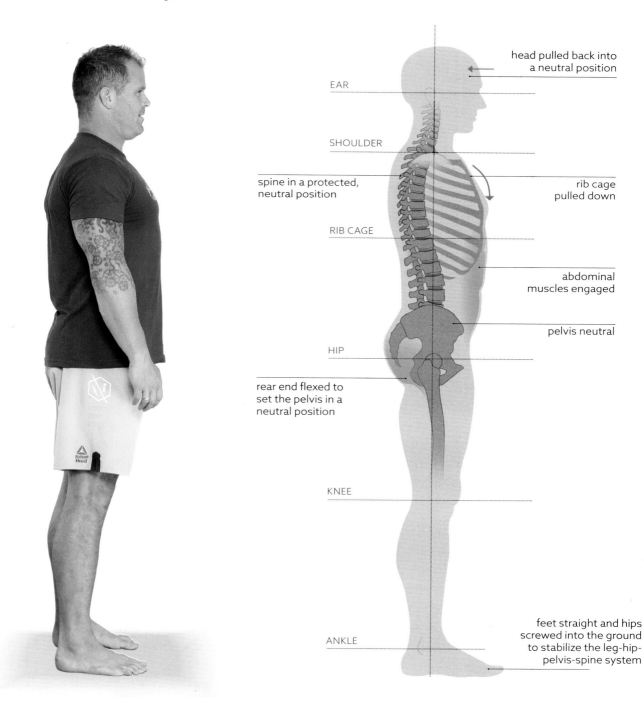

EAR

SHOULDER

spine in a protected,
neutral position

RIB CAGE

HIP

rear end flexed to
set the pelvis in a
neutral position

KNEE

ANKLE

head pulled back into
a neutral position

rib cage
pulled down

abdominal
muscles engaged

pelvis neutral

feet straight and hips
screwed into the ground
to stabilize the leg-hip-
pelvis-spine system

Before we get into the details of this foolproof bracing sequence, we want to explain the philosophy behind it. If you're not correctly engaging your myofascial system, the web of connective tissue and muscle that surrounds and supports your spine, you are either hanging on your soft tissues, discs, and ligaments (which occurs when you round forward in flexion) or creating bone-on-bone stability by slamming together the vertebral motion segments of your spine (which occurs when you overextend).

Some movement coaches and posture specialists will tell you that if you're in an organized position, you can completely relax. They argue that when the spine is perfectly aligned, with each vertebra stacked on top of the next, maintaining that position should require very little effort. This might be true if you're leaning back in a recliner, but 90 percent of daily activities involve movement. Even sitting at a desk involves movement. And if you're moving without employing these bracing tools, that perfectly organized spine will become disorganized and sloppy very quickly.

The only time you get to go on a spinal stability break is when you pass out at night during sleep. Otherwise, the rules always apply.

We have always believed that a lack of spinal integrity is one of the greatest limiting factors of human function, and not just for athletes. Maintaining the integrity of your spine can mean walking a 5K more quickly or picking up a heavy basket of laundry without fear of tweaking your back. No matter who we are working with, we start with the bracing sequence.

In case you're still not convinced, let us leave you with four important reasons to prioritize bracing your spine in a neutral position:

1. **It helps prevent insult and injury to your central nervous system.** This kind of injury can put your entire life on hold. It can affect your sleep, sex, work, and overall happiness. It's possible to rebound from some spinal injuries, but others bring about lifelong misery. We regularly see young people with spine-related injuries like fractures and disc injuries well before their 21st birthdays. The structures of the spine were intended to last a long time—certainly longer than two decades! The best way to prevent spinal injuries is—you guessed it—to learn how to brace your spine in a neutral position.

2. **It can increase your joint ranges of motion.** A disorganized spine can limit your ability to move freely because your central nervous system will protect itself by limiting force production and altering how your tissues adapt to movement. The nerves that exit your spine do not stretch, so when you create hinges, kinks, or instability in your spinal system, your body recognizes those as spinal liabilities and tightens the surrounding

musculature to limit your range of motion and thus reduce the potential for injury. For example, if you overextend through your lumbar spine, then your quadriceps, your hamstrings, and even your calves will tighten to protect your central nervous system. When you brace your spine in a neutral position, on the other hand, your muscles relax, giving you access to untapped mobility.

3. **It allows you to safely and effectively transition from one functional position to another.** You need to be able to change your position without compromising your spinal integrity. For example, think about how many times you rotate from side to side. Most people would agree that lifting something heavy with a rounded spine isn't a great practice. Add twisting to that scenario and you'll have to search long and hard before you find someone who will advocate for this compromised shape. We rarely point out a poorly organized spine in more pedestrian movements, such as getting into and out of a chair, but making these small, seemingly inconsequential movements with a poorly organized spine not only takes their toll over the years, but also ingrains poor movement patterns. So when you go to lift that heavy something and twist, you are going to do it with a spine that is not braced, which can lead to a spinal tweak.

If you do yoga, think about how many yoga poses require you to practice rotating in a controlled manner. It's like even the yogis of a thousand years ago knew something about the human condition.

4. **It helps you buffer bad working and life situations when you just can't be in an optimal shape.** For example, think of a police officer in body armor and utility belt sitting in a cruiser for eight hours. Or a harried mom reaching into the backseat to pull her crying infant out of the car seat. The braced neutral position is your base from which to generate all movement. However, we don't expect you to move like a robot and keep your back perfectly straight every time you change your position. On the contrary, organizing and stabilizing your spine will unlock all the capacity for incredible movement that your spine is designed for. The key is to have a base position, or starting shape, that is automatic so that you're not hammering your body with poor positions all day, every day. It keeps your body strong so that when you do lose focus and move incorrectly, you can handle the compromised motion. The last thing you should be thinking about as you move through the world is what your spine is doing. However, achieving this ultimate freedom requires some practice and some initial intention.

Creating a Ready Spine

The organizing and stabilizing sequence that we use is very straightforward and universal. We even use it with MLB pitchers, NFL linebackers, PGA golfers, and professional ballet dancers. Once you've practiced it a little, you can do it in a couple of seconds or less. The difficult part is becoming accustomed to it. As with anything in life, practice makes permanent.

The majority of us in this deskbound society have been practicing sitting, standing, and moving in less than optimal ways for a long time. The good news is that we are all capable of learning new tricks, especially when those tricks are ingrained in our DNA. It probably won't be as easy as it is to simply default to your old spinal habits. Your brain will likely be on overdrive as you get used to the steps and fight against the inefficient positioning that you have assumed for so many years. It might take you a few weeks of constantly checking and rechecking your posture until it becomes natural and ingrained. This is all right. Learning to move well is a skill, and it's one that is never finished, only refined. But just like learning to juggle, it will stick once you have it down.

Although there are only four steps, a lot is going on when you reset your spinal position. You have to recapture your pelvis, rib cage, and head alignment; you have to stabilize this position using the musculature of your trunk; and you have to organize your shoulders and hips to add crucial stability to the overall system. Trust us when we tell you that with practice, it will seem very natural. We've taught this sequence to literally tens of thousands of adults and children.

As with any difficult task, it is important to take baby steps. So, before we show you how to apply our bracing model from the standing position, let's take gravity out of the equation and create an organized spine while lying down.

Lie on the floor with your hands palm-up by your sides, then squeeze or activate your glute muscles.

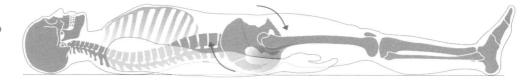

The reason for starting on your back is simple: lying on the ground removes complexity from the equation. When you're standing, you must contend with a number of forces that are working against gravity to keep you upright. While you are initially working on finding neutral, the familiar patterns of your body are doing what they always do—that is, pull you back to what you know. Working from the floor minimizes the adaptive loads on your spine. Your rib cage, head, and shoulders automatically assume a more neutral position because the ground is helping to align them.

Ready to try? Here's what we want you to do: Lie on the floor with your hands palm up by your sides. (If you're a yoga practitioner, you probably know this as the Shavasana pose.) Unless you have an injury, this position might be comfortable for a few minutes, but most people have a difficult time maintaining it for prolonged periods, especially on a hard floor. While lying with zero tension on a hard, flat surface, the body often defaults to a position of slight overextension, which for some people can register as low back pain or discomfort. This is why people tend to cross their legs while lying down. We will talk more in the next chapter about how rotation at the hip and shoulder joints contributes to the stability of the spine, but for the purpose of this exercise we don't want you crossing your legs.

The first step in the sequence is simply to squeeze your butt, or contract your glutes. You don't have to give it your all; just squeeze to about three-quarters of what your peak effort would be, or tightly enough to reorient your down-titled pelvis slightly upward toward your rib cage. After holding this position for a few moments, you should notice the tissues of your low back lengthening and any tension you were feeling in that area dissipating.

By engaging (squeezing) your glute muscles, you just realigned the relationship between your pelvis and your lumbar spine. The ground automatically put your head, shoulders, and rib cage into better alignment, and by flexing your glutes you pulled your pelvis into a naturally more organized position as well. Congratulations! You just found neutral, or home base—the position that reflects the normal and functional anatomical relationships of your spinal complex. Note that we don't talk about trying to tilt your pelvis or find some nebulous "perfect" pelvic orientation. Instead, we trust that your own personal anatomy will correspond to the musculature that accompanies it. Your glutes are designed specifically for your pelvis.

Don't think that this simple movement isn't sophisticated enough to capture the incredibly complex underlying physiology. Very much the contrary. Using the powerful muscles of your backside is a proven technique to ensure a stable and well-organized spine. Gymnastics coaches have been yelling, "Squeeze your butt!" to their athletes for as long as gymnastics has been around.

Two more thoughts:

- If you didn't feel your pelvis change positions when you engaged your glutes, don't worry; your pelvis was likely already in a good relationship with your lumbar spine.

- No, you won't have to walk around all day squeezing your glutes like a maniac. The glute squeeze simply resets your position and reclaims your body's correct geometry.

The next step is to brace, or set, this newfound spinal position by engaging the musculature of your trunk. Before we dive into that, though, we need to take a moment to talk about breathing mechanics. Of all the incredible functions that our bodies perform, breathing is one of the most overlooked and undervalued. We just had you reset the position of your pelvis so that we could begin talking about how breathing and spinal stability go hand in hand. And remember what we said about a disorganized spine: it causes many functions of the body's tissues to become positionally inhibited. This is especially true for your diaphragm, which is the heart of your beautiful breathing machine. Now that you've created a better relationship between your pelvis and spine, we can get to the interesting conversations.

The Lost Art of Good Breathing Mechanics

Breathing, like moving, is innate and vital, so we don't typically think of it as a developable skill that merits our attention. But just as the quality of the body's organization as we stand, sit, and move ranges from poor to optimal, so does the quality of our breathing. The problem with being a modern human is that our environment can exert significant forces on our bodies that cause us to become maladapted. Remember our example of the heel-striking first graders (see page 13)? Well, most of us are doing the equivalent of heel-striking when we breathe. And remember when we said that we start all of our conversations with spinal mechanics? Well, we can't talk about spines and not also talk about breathing. It's like forgetting that your head sits on your neck. They form an integrated system.

In fact, we can take this relationship a step further and expand our definition of spinal organization to include the central nervous system. (It's not a very big step, actually.) How you breathe—whether it is shallow neck breathing or powerful breathing initiated by the diaphragm—affects the ways in which your body interprets stressors. In short, poor spinal mechanics changes the way you breathe (see "Diaphragm Dysfunction," beginning on page 45).

Your body interprets shallow neck breathing—which is what we observe in typical sitting populations as well as those who are exercising and out of breath—as a cue to fire up your fight-or-flight chemistry. That ability to recognize stress breathing and react with the appropriate stress hormones is a useful adaptation when you're chasing dinner prey or defending your babies against lions. But just as drinking coffee all day long impairs your ability to wind down and go to sleep, so too does the stress signal that you are sending to your brain throughout the day by neck breathing. This is one of the underlying themes in all the research that implicates sitting in wrecking our physiology.

The first step in breaking this stressed breathing pattern is simply to align your spinal breathing carriage into a better shape. When you're in a neutral position, it is typically easier for your body to default to its natural and efficient breathing pattern.

Lie on your back, just as you did for the previous exercise. But this time, we don't want you to squeeze your glutes or try to optimize your alignment. In fact, bend your legs and bring your heels toward your hips. This legs-bent position removes tension from your trunk and diaphragm by creating slack in your lower extremities. When you are lying down, your trunk musculature doesn't have to support your body in gravity, which makes breathing into your belly less complicated.

Next, place your hands on your belly, one on top of the other. Then take a slow, steady breath in through your nose, directing the air into your belly. Think about lifting your hands with your stomach as you breathe in. When you do it correctly, your chest will remain still, your belly will swell, and your hands will rise.

Don't worry about taking a deep breath or emptying all the air out of your lungs. Just breathe as you normally would, inhaling and exhaling through your nose and belly. If possible, put down this book and perform this exercise for two minutes. Your goal is effortless belly-based breathing.

Let's return to the coffee shop human observatory for a minute. Last time you had to do an espresso shot for every rounded spine you saw. This time, secretly observe how people are breathing. Buy everyone a venti decaf hazelnut vanilla soy latte if you can spot even one person using her diaphragm to perform belly-based breathing. Don't worry, you won't have to. When we sit, our breathing mechanics tend to go out the door. The real problem with our hijacked breathing plan is that we are practicing this dysfunction for most of the 20,000 breaths we take every day.

Belly-based breathing is the way you want to breathe when at rest and while performing normal day-to-day activities. Breathing diaphragmatically grants you access to your parasympathetic nervous system, which does the opposite of your sympathetic, fight-or-flight nervous system. Wonder why we say "take a deep breath" to someone in a stressful situation? Yep, you

Diaphragmatic Breathing

1. Lie on the ground and place your hands on your belly, one on top of the other. To reduce tension in your trunk and low back, bend your legs and position your heels near your hips.

2. Take a slow and steady breath in through your nose, directing the air into your belly.

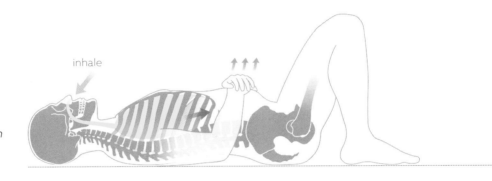

3. As you exhale, your hands should drop. The idea is to move your hands up and down using your breath. This is how you breathe diaphragmatically.

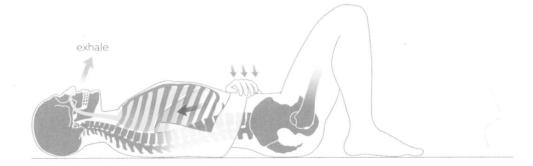

nailed it. Deep breathing, aka belly breathing, is a direct shortcut to the ability to down-regulate or de-stress. Unless you're doing something strenuous, like running or lifting weights, you should be breathing in and out through your nose and belly most of the time.

You may be wondering why else diaphragmatic breathing is beneficial. Well, for one, breathing in through your nose relaxes the muscles of your neck, face, and jaw. Ever wonder why running coaches cue their athletes to relax their faces while running fast and, by extension, breathing hard? What's more, the airways in your nose are a lot smaller than the airway in your mouth. In addition to allowing you to avoid being labeled a "mouth breather," breathing through your nose creates resistance for your diaphragm to pull against, helping you take longer, deeper breaths.

To be clear, your neck, chest, back, and rib cage all expand when you take a deep breath. For example, if you're getting ready to hold your breath underwater, you want to take in as much air as possible. You do so not only by breathing in through your belly, but also by expanding your neck, chest, and back to reach maximum lung capacity. For everyday, run-of-the-mill breathing, think about starting your breath in your belly. This breath then carries up the sides of your rib cage. Imagine that you are an Oreo cookie and that breathing into the sides of your rib cage is like breathing into the cream filling between the two cookies.

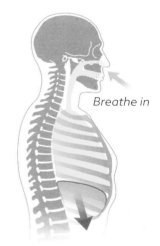

Breathe in

When you're working at a computer, answering emails, or just going about your day, breathe diaphragmatically. If you're stressed out or you find yourself defaulting to your old, compromised breathing pattern, direct your attention to your breath. You'll probably notice that you're breathing into your chest and neck. When this happens, take a two-minute break and get your breathing in check.

The easiest way to tell your body that nothing is wrong and that there are no imminent threats to your survival is to practice your diaphragmatic breathing. The best part is, you can do it anywhere. You don't have to lie down, although lying down is a nice way to relax and ensure that your spine is in proper alignment as well. You can do it while sitting, standing, or even walking. Nobody has to know that you're doing a breathing exercise. Simply place your hands on your belly and breathe in and out through your nose, directing the air into your belly and up the sides of your rib cage.

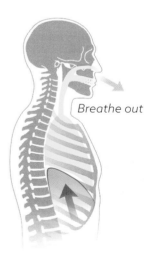

Breathe out

Box Breathing

If you're feeling anxious or panicked, it can be difficult to access your diaphragmatic breathing, especially if you're hyperventilating—breathing in rapidly through your mouth and chest. In situations of extreme stress, we recommend a breathing exercise called box breathing. Former Navy SEAL Mark Divine—the founder of SEALFIT—teaches this technique to all of his tactical athletes and war fighters. He uses it to help his soldiers get their breathing back under control when they are feeling stressed. If you find yourself in a highly stressed state, take two minutes and perform this drill. But it's important to point out that this is a breathing drill, not an everyday breathing pattern. Here's how it's done:

1. *Breathe in through your nose and into your belly, taking in as much air as you can. It should take you at least four to six seconds to inhale.*
2. *With your lungs full of air, hold your breath for two seconds.*
3. *Exhale slowly, taking a full four to six seconds to expel all the air out of your lungs.*
4. *With your lungs completely empty, hold your breath for two seconds.*
5. *Repeat for at least two minutes.*

Stabilizing Your Spine (and Breathing at the Same Time)

Now that you are able to find an organized spinal shape and you understand how to breathe through your diaphragm, the next step is to stabilize, or brace, this position.

Lying on your back in the same position as before, squeeze your glutes to reset your pelvis-lumbar relationship. Then take a big breath into your belly. As you exhale, create tension in your abdominals by trying to get your abs to drop flat to your spine, as if you were trying to shrink-wrap your trunk around your spine. Don't suck in or hollow out your belly while you do this; that it is a terrible way to move and function and becomes a silly long-term strategy as soon as you need to breathe again. "Suck it in" is the worst advice you could give your poor spine. You shouldn't see your abdominals bulge out, either. Instead, think about creating stiffness around your spine. We think this is easiest to learn on the exhale. Drawing your belly button away from your pants on the exhale helps you visualize that you are trying to create the correct resting posture for your trunk muscles.

A smaller space around your spine is easier to stabilize and is vital for the functioning of a healthy nervous system. The higher intra-abdominal pressure created by this stabilization helps keep the vertebral discs from carrying all of the load, all of the time.

The next step in the process is to grade the amount of tension that you are creating around your spine. Whenever you are sitting upright or standing, your trunk musculature should be engaged and supporting your spine. As we've said before, the only time you get to turn off your primary spinal stabilization mechanism is when you are reclining (see pages 187 and 188), lying down, or asleep. Otherwise, your trunk really doesn't get to go on vacation. To simplify this concept, we recommend that you maintain a default working tension of 20 percent.

How do you judge 20 percent? Unfortunately, it is not an exact science. The best way to find the right amount of tension is first to relax completely; that represents zero tension. Then, on the exhale, stiffen your trunk as hard as you can; that is 100 percent. Finally, use your best judgment to scale it back to 20 percent.

Maintaining this amount of tension throughout the day is quite manageable, and it puts you in a ready state to quickly increase your spinal stability when needed. This background stabilization program allows your body to optimize the reflexive spinal stability that is part of your hard-wiring. Instead of having to go through the bracing sequence in a split second, you are already there. All you need to do is turn that 20 percent up to 100 percent. Think of it like driving a car: it's much more efficient to go from 20 mph to 60 mph than it is to hit 60 mph from a dead stop.

Maintaining a constant 20 percent abdominal tension gives you inertia so that you can ramp up to meet the demands of your next movement. Most of us are good at "getting tight" for a few seconds if we know what's coming. Unfortunately, this isn't the way the world works. Your goal is to create a running stabilization that you don't have to consciously attend to very much. You need to be able to quickly reach into the crib and grab your crying baby without having to get your spine organized from scratch.

You should always be protected by your spinal stabilization program. It may seem unfair that you need to practice good spinal hygiene, but think of it like brushing your teeth before you go to bed—eventually, you won't have to think about your spinal hygiene, either. Moving well is a skill that is easy to practice. But like any new skill, it has to be practiced before it becomes ingrained.

We're not going to lie; creating and maintaining default abdominal tension seems like work in the beginning. You're engaging muscles and tissue systems that may have been asleep for years. But believe us when we tell you that it will become instinctual with practice. How long will it take? It depends. Some people pick it up within a week, and for others it takes months. The bottom line is that you can't get around the need to have a cogent, reproducible plan for stabilizing your trunk. If you've ever taken a Pilates or yoga class, you already know that human beings have acknowledged this need for a very long time and have thought critically about how to teach it.

Now that you know what it feels like to create the tension necessary for routine activities, let's see what it feels like to create maximum tension. To perform this drill, we want you to imagine that someone is standing above you, about to drop a bowling ball on your stomach. If your trunk is relaxed and you have a belly full of air, what do you think will happen on impact? Not pretty, right? In order to absorb the impact of the bowling ball, you will need to quickly let out all of your air as you tighten your trunk to 100 percent tension.

When you ratchet up to 100 percent, you will soon become aware that it is much more difficult to breathe through your belly than it is at 20 percent. That's the point: you've created a sort of "cast" around your spine. If you're trying to break the deadlift world record, you don't need to breathe during the movement. Similarly, if someone is about to punch you, you don't need to breathe at the moment you take the hit. Instead, you need to create as much tension around your midsection as possible to protect your spine from the blow.

The hard part is cycling in and out of peak tension while maintaining a neutral spine and a diaphragmatic breathing pattern, as well as judging how much tension you need at any given moment. Even cobras don't flex their hoods all the time. How much tension is required to run a 5K? More than you need to walk to the mailbox, but less than you need to pick up a heavy-ish box. Failure to stabilize and breathe is a common problem that we see even in our top athletes. Have you ever seen the strongmen at the base of the pyramid at a Cirque du Soleil performance? They work very hard to support the load of multiple athletes for sustained periods and breathe at the same time. What you may not understand is how difficult it is to maintain near-maximum trunk stiffness under high aerobic demands.

We've covered a lot of ground here, so let's recap quickly: your glutes put your pelvis in alignment with your lumbar spine and rib cage (thus getting your spine organized), and you lock everything in place (or stabilize the position) by using your abdominal muscles. Simple, right? Before you go out and throw a party to show off your shiny new posture, remember that you have one more element to factor in, which is how to breathe in that new organized and stabilized posture.

The Aerobic Spine

When people don't have a strategy for bracing and breathing at the same time, they often have to choose one or the other: either they stabilize by holding their breath, or they sacrifice spinal stability and breathe. There is a better way.

Here's what we want you to do: Lie down on your back, engage your glutes to set your pelvis, tighten your abs slightly to lock everything in place—creating about 20 percent trunk tension—and then place one hand on your stomach and the other on your chest. Now breathe. Again, imagine pulling air into your belly rather than into your chest. The hand on your stomach, not the hand on your chest, should move up and down as you breathe. Our mission here is to get you to realize that you *can* breathe through your stomach with your abdominal muscles turned on. Just because they are slightly flexed doesn't mean that they can't expand and contract. If you need to create more tension, you're ready. All you have to do is let out some air while simultaneously creating additional tension in your trunk.

This simple drill even scales to Olympic-level loading. One of the cues that we give to our strength athletes is to fill their bellies with air before they lift. Given what you now know, this cue makes good sense. Higher intra-abdominal pressure means a more stable spine. However, athletes often mistakenly apply this cue when they fail to stiffen their trunks prior to packing the air in. You can see this for yourself. Relax your belly and take in as big a breath as you can. Now try to stiffen your trunk around that pillow of air. Simply put, you can't. You must first create a steel-like cylinder out of your trunk tissues and then try to get as much air into this confined space as possible. Are you about to lift something really heavy? This is your plan.

We know that this sequence of drills can seem laborious, but mastering breathing through your diaphragm while keeping your trunk tensed is not a step that you want to skip. It's a crucial component of the bracing sequence because in order to stabilize your spine, you must shrink the space around it by using the musculature of your trunk. You've got your trunk and diaphragm set, and when you reset your pelvis-lumbar relationship with the glute squeeze, you've even set the conditions for your pelvic floor to form the bottom of your spinal breathing system.

Thus far we've been talking about the model for organizing and bracing your spine from head to pelvis. If we stopped here, we'd be making real progress, but as it turns out, you have arms and legs, too. And believe it or not, your hips and shoulders play a crucial role in creating a stable spine. In fact, failure to account for hip and shoulder mechanics is precisely why so many people cannot transfer their carefully cultivated spinal mechanics to the rigors of the real world.

The Rotational Key: Stabilizing Your Hips and Shoulders

A well-organized and stable spine is truly an evolutionary wonder to behold. Now imagine that beautiful machine trying to balance itself on top of your leg bones (your femurs). It would be a wobbly mess indeed. This isn't what happens when you are standing, of course, because your body makes use of a highly effective stabilizing force generated by a specific rotation (twisting) of your anatomy. You are actually already familiar with the concept. If you lie on your back and relax, you'll notice that your feet rotate away from one another like two repelling magnets. What's happening is that the natural winding mechanism that your body employs to create stability from your feet to your hips is unwinding—just like a twisted rubber band unwinds when you let go of one side. Your legs are naturally endowed with tissues that become stable when a rotation force is applied.

How do you apply that rotation force? You create an organized pelvis and stand with your feet straight. Situated at the interface of the hip and pelvis, the head of your femur looks a lot like a ball wrapped in connective tissue (the connective tissue capsule of the hip joint). To secure a piece of candy in a wrapper, you simply twist the ends of the wrapper. The same applies to hip architecture. Rotation (the "twisting of the wrapper") is how our bodies create stability at both the hip and the shoulder. And since the design of the hip joint is remarkably similar to that of the shoulder joint, it's no accident that evolution settled on the same stability solution for both structures: rotation.

If you've ever been around a serious sport, you may have noticed that most of the cues that coaches deliver fall into one of two categories: they are either cues about creating a stable trunk or cues about creating rotation. Humans are clever and performance-obsessed enough to have developed movement techniques that optimize our anatomy.

The good news for us is that we can translate the old movement language and teachings of generations past into practical guidelines for how to improve our own function, even while working at a desk.

Rotational Stability Through the Hips

Remember that old song about how the bones of the body are connected? That song is a pretty powerful allegory for the connected functional nature of our bodies' various components. A stable, organized spine and pelvis sets the conditions for hips that are able to work at full capacity. It should come as no surprise, then, that stable hip joints reciprocate stability back to the pelvis. What we are saying is that, in order to complete the spinal stability model that we've been developing, we have to include some basic hip function and organization. You'll never have a rock-solid spine if the downstream connections of that system (your pelvis/femur joints) are not accounted for.

This all sounds a bit complicated, but putting it into practice is easy. Here's your first exercise.

Stand with your feet underneath your hips, take yourself through the bracing sequence (you'll find the steps on pages 82 and 83), and then screw your hips into the ground without moving your feet. Your right hip will rotate in place in a clockwise direction, and your left hip will rotate in place in a counterclockwise direction. Imagine that your feet are on dinner plates and you are trying to rotate them away from one another. (*Note:* Your feet don't actually turn outward when stabilizing your hips—see page 76—you're just creating an outward force.) The twisting force that you generate here needs to be initiated in your hips, not at your feet. Remember, good movement starts at the spine and moves outward. This rotation force is the movement key that ties the whole leg to the pelvis and spine.

Notice what happens in your lumbar spine while you are screwing your hips into the ground. You'll probably find the tension in your lumbar spine and pelvis area lessening. This happens because the rotation force that you are generating in your hips works to support a well-organized pelvis. This external rotation force reorients your pelvis into a more ideal shape, which makes it easier for your stabilizing trunk and gluteal musculature to keep your trunk upright and organized on your femurs.

With hip rotation now a part of your organized body, you can add another 20 percent force to the mix. We like to advocate for 20 percent tension in the glutes, abs, and now rotation through the hips. If you've ever been to a yoga class, you might recognize this hip-to-foot rotation as part of Tadasana, or Mountain Pose. Not only does this external rotation force through the hips stabilize the hip joints and spine, but it is also one of the ways in which the body creates stable feet and ankles. In fact, if you notice that your ankle bones aren't squarely in the middle of your feet, you may need to create a little more rotation at the hip. A large component of creating a foot with a solid arch is generating rotation in the hip that extends through the rest of the leg.

Rotation Force: Hips

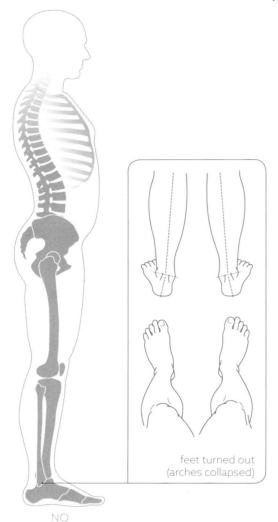

feet turned out
(arches collapsed)

NO
ROTATION

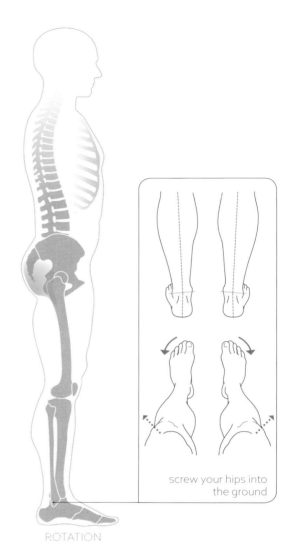

screw your hips into
the ground

ROTATION

When we teach seminars, we regularly see rooms full of collapsed arches and turned-out feet. Most of us treat the ligaments of our feet as seat belts, passively leaning against them without giving them much thought. Your feet are little miracles of design that utilize muscles, connective tissues, and ligaments to maintain their shape. Failing to actively support your feet is a model for long-term system collapse.

Having trained tens of thousands of coaches and athletes in person at our seminars, we've never met an arch that we couldn't reclaim. Try it for yourself: Step 1 is to stand with your feet pointing straight forward, and Step 2 is to screw your feet into the ground from the hips—remember, right foot clockwise, left foot counterclockwise—after you've organized and braced your spine. Now try to collapse your feet. See? You can't. That's the point.

You might be asking, Can't I create the same amount of hip rotation even if I'm not standing with my feet pointed straight ahead? The short answer is no. When your feet are turned out much past 12 to 15 degrees, it's much harder for your hips to create a rotation force. Don't take our word for it; experience it for yourself. Generate as much external rotation force as you can from your hips with your feet pointed forward, and then try to match that amount of force with your feet turned out 30 degrees. You'll feel much weaker with your feet turned out.

Remember, your body always functions better when it is organized. And besides, we are talking about long days here. You want the biggest payoff for the least amount of effort.

Rotational Stability Through the Shoulders

Now that you understand the organizing principles for the hips and spine, let's take this concept upstream to your shoulders. Like your hips, your shoulders are ball-and-socket joints. In fact, when it comes to creating stability, it's helpful to think of your hips and shoulders as the same kind of joint, because the technique for creating stability while standing and performing the vast majority of movements is the same. And just as lower spine stability is incomplete without the stabilizing rotation force of the hips, so too is the stability of the neck and rib cage incomplete without well-stabilized shoulders. What this means is that head and neck–related problems of position cannot be fully remedied without also addressing the mechanics and positioning of the shoulders. It should be obvious on the surface that your neck and shoulders share common tissues and structures. But our reductive thinking has tricked us into forgetting that these structures are symbiotic. When physical therapists assess people for neck or shoulder pain, they also assess the other structure. Neither gets a clean bill of health until both do.

To create functionally stable shoulders, you need to apply the same rules you used at the hip complex. What happens when you fail to stabilize your shoulders in a well-organized, externally rotated position? Your shoulders roll forward, internally rotating. This pulls your upper back into a rounded shape and forces your head out in front of your body. Think critically for a moment: how many of us have shoulders that stick out in front of our chests when we are sitting or standing? The answer should be "none of us," but in reality it is "nearly all of us."

Let's test this principle to see how it works in practice. Stand with your arms outstretched at shoulder height—forming the letter *T* with your body.

Rotation Force: Shoulders

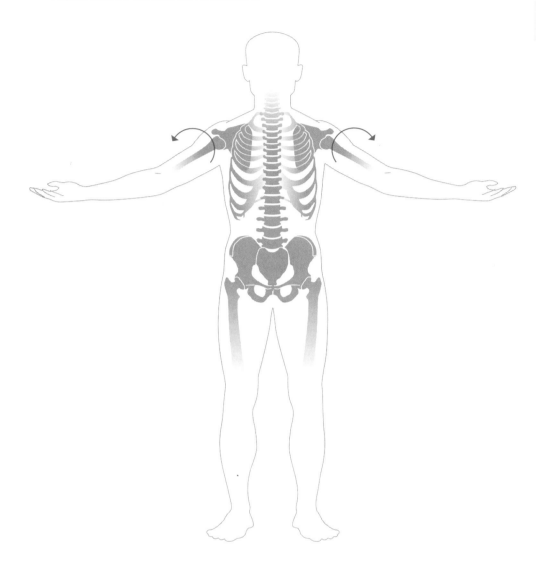

To stabilize your shoulders in a neutral, externally rotated position, stand with your arms at your sides, raise them to chest height—keeping your arms aligned with your body and your spine neutral—and flip your palms to the ceiling.

Next, turn your palms toward the ground. Notice how this rotates your armpits downward. Most of us will simultaneously experience a rolling-forward (or "forward translation") of our shoulders. If your shoulders rolled forward instead of spinning in place, you experienced one of the root causes of dysfunction in the neck-shoulder-arm-wrist system.

Now let's correct this inefficiency and get your shoulders into a more sustainable shape. Stand with your arms at your sides, raise them again to chest height, flip your palms to the sky, and try to get the pits of your elbows to face the ceiling. The external rotational motion that you have generated has wound up your shoulders in their capsules, reset your scapulae into a good position with your ribs, and moved the heads of your arm bones into better position within the joint sockets.

Do your shoulders feel more stable? We suspect that the answer is yes. We'd be willing to bet that your head automatically got pulled back into a better position, too. Now relax into your old pattern. Do your shoulders roll forward? Do your thumbs point at each other instead of pointing straight ahead? See if you can create the same organizing rotation with your hands at your sides. Remember, the goal is to rotate your shoulders into stability, not simply to twist your hands to make it *look* like you are doing it.

When we ask people what shapes their shoulders should be in, most folks talk about trying to adjust their shoulder blades, and when they try to stabilize their shoulders, they perform some kind of weird shoulder/scapula shrug "back and down." This effort may move the shoulder blades into a slightly better shape, but it fails to address the relationship of the arm to the shoulder. What you get are the same poorly positioned shoulders and a jutted-out rib cage. Instead, focus on externally rotating your shoulders deep within the joints. Your pectoral muscles should be behind your shoulders, and if your hands are at your sides, your thumbs should be pointing forward. Don't worry; the scapulae will automatically orient themselves without any additional action on your part.

When Leonardo da Vinci drew the anatomical man with palms facing forward, he was showing us how our shoulders should be organized. He was giving us clues about how to create a stable shoulder-neck system. Once you've organized your shoulders in a stable spinal carriage, the new goal is to maintain the integrity of this orientation independently of what your arms are doing. In other words, you can position your arms however you like— whether you're typing on a computer, talking on the phone, or just standing around—but your shoulder position should remain consistent.

It's important to note that when your shoulders are well organized, the musculature responsible for orienting and stabilizing your shoulder blades becomes much more active. When people are first turned on to good shoulder positioning, they often report feeling fatigue in the musculature that supports their shoulder blades. This kind of fatigue is good. It means that your structure is finally doing its job! It may take some time (not a lot, mind you) to develop and condition these muscles, but stick with it. As we've said, your muscles and tissues are like obedient dogs. Once they're strong enough to hold your new posture, it won't feel like work at all. It will feel natural and relaxed.

In Section 3, we show you how to apply these organizing principles for the strong engines of your hips and shoulders to fundamental human movements like hinging, squatting, and pressing. But first, let's take all the concepts we've covered and combine them into one simple, seamless sequence.

The Bracing Sequence: Reclaiming a Good Spinal Shape

Most of the people we work with recognize the importance of an organized and stabilized spine, but if we ask them to take us through the sequence of steps to do it, we usually get blank stares in return. We even get this response from people who participate in yoga, Pilates, martial arts, gymnastics, and other practices that prioritize aspects of spinal alignment. When we ask these people to brace their spines, many of them do a damn decent job. The breakdown occurs when we ask them to take us step by step through the process. Nine times out of ten they will say something like, "Steps? I don't have any steps; I just kind of do it." Or, "You know, get tight."

Why is this an issue? Because no matter how good you are at stabilizing your spine in your sport or practice of choice, eventually you will find yourself in an unfamiliar situation or position. And, as we all know, injuries and body tweaks don't always occur while you are doing something dramatic, like lifting a heavy weight over your head. The spine is more vulnerable when left unprotected. You need an observable, measurable, and repeatable model that works for every age, task, and situation.

Any system worth its merit must be scalable and easily transferable—it must be as easily grasped by a desk jockey as a world-class athlete; it must be the same for striking a yoga pose, fighting in the ring, or bending over to pick up your keys; and it must give you the same results whether you are moving fast or slow, hefting something heavy or throwing something light. You need a step-by-step process, not a meaningless catchphrase such as "get tight" or "strengthen your core."

The bracing sequence is that system. This blueprint will give you the same dependable results every time.

The goal is to make the bracing sequence a part of your everyday routine. Every time you catch yourself out of position—say, while working at your computer or just standing around—stop what you're doing and reset your position by going through this simple process.

We know that keeping a portion of your attention on your posture throughout the day might seem like a chore at first. But rest assured that with practice, maintaining an organized body will become instinctual. As your posture improves, you won't have to go through the bracing sequence to correct your position nearly as often, because you'll already be in a good position.

While it may seem like we've covered significant ground just so we can more easily transition from sitting to standing, the principles that we've covered in this section are the foundational language of human movement. Most of us want to get right to writing poetry, but we don't yet have the basics of grammar, spelling, and punctuation down. Now you do. Mr. Miyagi of *Karate Kid* fame had it right: "First learn walk, then learn run."

The Bracing Sequence

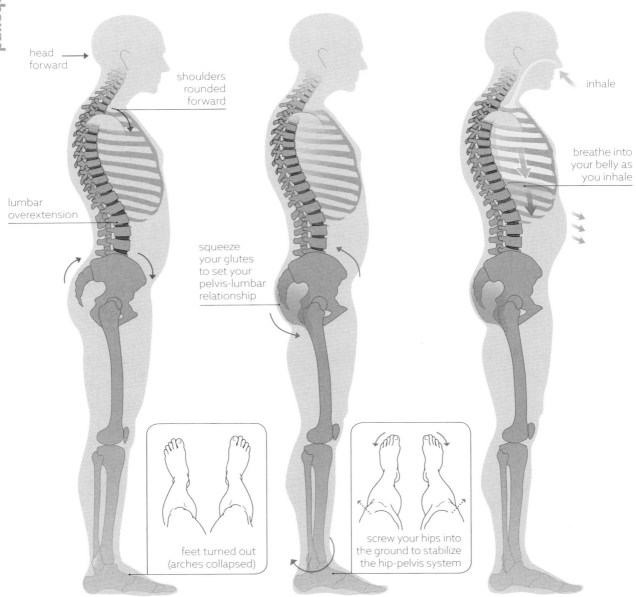

head forward

shoulders rounded forward

inhale

breathe into your belly as you inhale

lumbar overextension

squeeze your glutes to set your pelvis-lumbar relationship

feet turned out (arches collapsed)

screw your hips into the ground to stabilize the hip-pelvis system

START
Compromised
Position

Step 1: Set your pelvis in a neutral position.
Position your feet directly underneath your hips with your feet parallel to each other. Squeeze your glutes to reset your pelvic orientation. Screw your hips into the ground, your left leg in a counterclockwise direction and your right leg in a clockwise direction. You're not letting your feet turn outward; you're just exerting a screwing-into-the-ground force while keeping your feet straight. You don't need to keep your butt at full tension; you need just enough glute engagement to orient your pelvis in a neutral position.

Step 2: Balance your rib cage over your pelvis and brace the position.
2.1. Keeping your glutes engaged and maintaining an external rotation force through your hips, take a deep breath in through your diaphragm (belly).

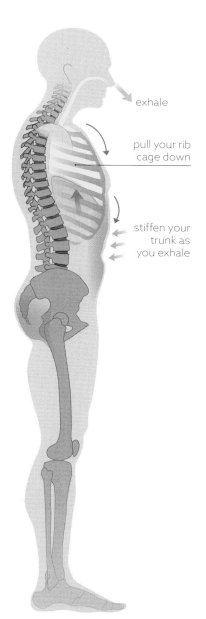

exhale

pull your rib
cage down

stiffen your
trunk as
you exhale

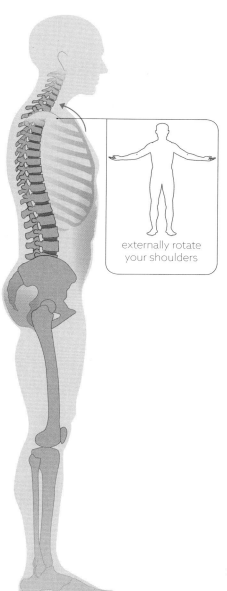

externally rotate
your shoulders

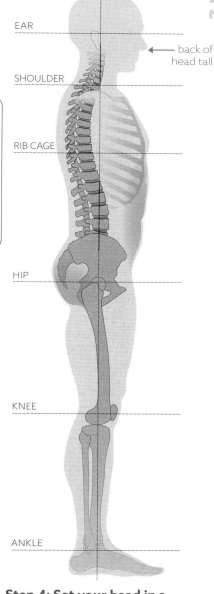

EAR

back of
head tall

SHOULDER

RIB CAGE

HIP

KNEE

ANKLE

2.2. As you exhale, orient your rib cage over your pelvis and pull your belly away from your pants (don't suck in or "hollow"), stiffening as you exhale. Remember, your glutes and hips set the position, and the muscles of your trunk brace the position.

Step 3: Organize your shoulders.
Keeping your arms positioned at your sides, screw your shoulders backward until your palms face forward. You want to draw the heads of your arm bones back, spreading your collarbones wide as you externally rotate your shoulders. Don't let your rib cage flare or tilt.

Step 4: Set your head in a neutral position.
Balance your head over your shoulders, focusing your gaze straight ahead. The goal is to align your ears over the center of your shoulders, hips, and ankles. To complete the sequence, let your forearms and hands relax at your sides, keeping your thumbs pointed forward and your shoulders in their casually wound-up, stable shape.

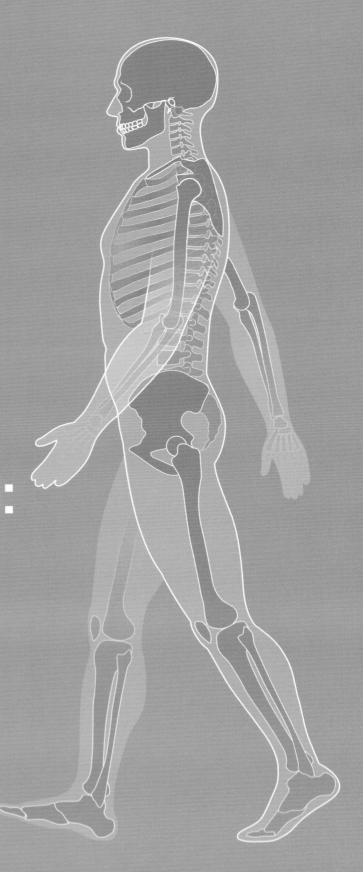

Moving Well: Walking, Hinging, Squatting, and Stable Shoulders

The biggest problem with the fact that human beings are such capable movers is that, on the surface, there appears to be no common thread running through the everyday movements that we perform. But once we overlay a basic schema onto the collection of wildly diverse and seemingly disparate movements that we naturally express, the human movement language becomes infinitely easier to digest.

Now that we've run you through the process of organizing and stabilizing your spine, you have the most important building block underlying nearly every movement that you might express in the course of a typical day, and then some. For example, let's look at sitting down: maintain your basic spinal shape, hinge at the waist, and lower your center of gravity. A push-up is just a neutral spine appropriately braced as you push away from the ground. Running, walking, squatting, and picking something up off the floor all involve the same spinal shape that you've already practiced. Think of it this way: the bulk of the movements that we perform are really just iterations of moving our arms and legs while maintaining a neutral spinal position. Not only does this simplify things considerably, but it also offers us a blueprint for spotting some of the faults underlying potentially harmful body mechanics.

You might conceptualize your spine—from head to pelvis—as a fence post or rod. It doesn't matter what you are doing (with the exception of gymnastics and tumbling movements); your spine doesn't really change shape. The complexity, speed, or load of the movement doesn't matter, either; when distilled, most movements express shapes that are nearly identical in organization.

The goal is simple: learn how to walk, hinge, squat, and stabilize your shoulders with an organized spine and then apply those principles to all real-world tasks. Achieving that goal will not only eliminate many of the problems associated with sitting and desk work, but also dramatically lessen your chances of suffering down the road. This section provides you with all the tools you'll need to accomplish this goal.

Walking

Walking is one of the safest and easiest ways to add movement and increase your non-exercise activity (see page 18) throughout the day. With the typical recommendation being to take at least 10,000 steps a day, it's in your best interests to ensure that you are walking well. Your lower back shouldn't hurt every time you walk more than a mile, and you shouldn't get bunions or weird calluses on your feet. These are symptoms of poor walking technique, tissue restrictions, or both.

The good news is that improving your walking is easy. In fact, you can dramatically upgrade your walking mechanics by following a few simple guidelines.

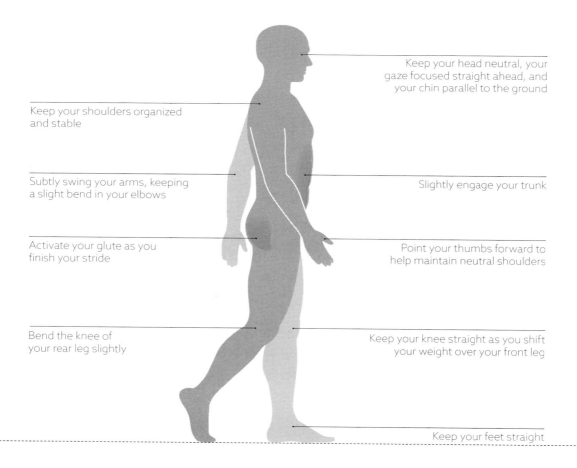

Keep your shoulders organized and stable

Subtly swing your arms, keeping a slight bend in your elbows

Activate your glute as you finish your stride

Bend the knee of your rear leg slightly

Keep your head neutral, your gaze focused straight ahead, and your chin parallel to the ground

Slightly engage your trunk

Point your thumbs forward to help maintain neutral shoulders

Keep your knee straight as you shift your weight over your front leg

Keep your feet straight

Walking Mechanics

Step 1: Stance and setup.

If you're wearing shoes, take them off. Position your feet directly underneath you, about a fist-width apart. The toes on both feet should be pointing straight ahead. Then go through the bracing sequence (pages 82 and 83).

Step 2: Step and shift your weight onto your front leg as it passes underneath your hips.

One of the most common faults associated with walking is stepping too far out in front of the body, or over-striding. Taking too long of a stride is like putting on the brakes every time you step. The key to walking efficiently is to step just in front of your body and shift your weight onto your leg as it passes underneath your hips. It's better to take shorter, more frequent steps than to take too long of a stride.

To help you understand how this works, try this drill (shown on the following page): fall forward from your braced neutral stance without looking down or hinging at the hips. The goal is to fall forward from your ankles as if they were your only joints and then allow yourself to take a couple of steps.

If you're like most people, you will stop yourself from face-planting by stepping your dominant foot forward. Instead of over-striding and distributing your weight over your leg while it's positioned out in front of your body (as you might do when walking in highly cushioned shoes), your foot makes contact with the ground just a few inches in front of your body, and then you shift your weight onto that leg as it passes underneath your hips, or center of mass. So you fall forward, step your foot to restore your balance, and then catch your weight. If you watch small children walk or run, they actually "fall" to initiate forward motion.

Good walking is actually a much more subtle version of this falling drill.

CORRECT

INCORRECT

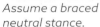

Assume a braced neutral stance.

Fall forward from the ankles.

Step your foot just out in front of your body to restore your balance.

Shift your weight onto your leg as it passes underneath your hips.

Another crucial point is how your foot strikes the ground. Over-striding occurs when you step too far out in front of your body, usually due to missing ankle and hip range of motion. For example, if you're missing hip extension range of motion (think of your leg moving behind your body), you will typically compensate by stepping too far forward. What you can't finish on the back end, you will make up for on the front end. Couple this tendency with extremely cushioned shoes and you end up with an exaggerated heel-strike. Take away your shoes and this destructive walking pattern would not survive. The impact on your heels would be too great. Your body literally won't allow you to do it for very long.

Try this experiment: while wearing your favorite sneakers, video yourself walking 10 to 15 meters at normal speed. Now repeat the process barefoot. What did you notice?

We bet that the way you walked changed dramatically. Without shoes, you probably took shorter and more frequent steps, initiated foot contact with your lead leg much closer to your center of mass, and ceased to slam your heels into the ground like a maniac. Now here's the rub: your walking should be mechanically the same, regardless of your footwear. The key is to roll your foot from heel to toe, like this:

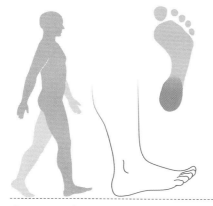 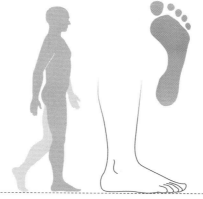 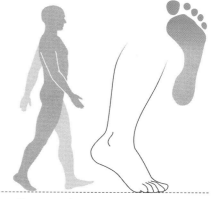

Contact Phase
You make contact with the ground with your heel during the contact phase. Remember, your foot should strike the ground just in front of your body.

Midstance Phase
As your back foot propels you forward, your entire lead foot should be flush with the ground, positioned underneath your hips, with the majority of your weight distributed over your grounded leg.

Terminal Stance
As you swing your opposite leg forward, naturally roll onto the ball of your foot. During this phase, you should feel your glute subtly activate as you fall into your next step.

Step 3: Maintain straight feet.

For this step you'll need a straight line on the ground to use as a reference. You could use a grout line on a tile floor or put down a strip of tape. Position your feet on either side of the line, again about a fist-width apart, and go through the bracing sequence. Next, perform the same falling drill from Step 2, but this time pay attention to how your foot lands. If your walk is mechanically correct, your foot should be the same distance from the line as it was before you took a step, and it should remain parallel to the line. If your toes are pointing out to the side, perform this step again, this time concentrating on landing with your foot straighter.

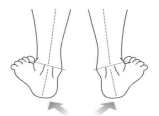

duck walk/stance fault (see pages 92 and 93 for more on this fault and how to correct it)

Walking Faults and Corrections

If you're free of negative adaptations and range of motion restrictions, correcting the way you walk by following the three simple steps we just outlined is relatively easy. All it takes is practice. However, a large portion of modern humans are simply unable to walk correctly due to a lifetime of poor posture. While you can use the mobility techniques laid out in Section 7 to eliminate these negative adaptations and restrictions, it is important to understand the greatest impediments to walking technique. Correcting the way you sit and stand will help move you in the right direction, but when it comes to walking, there are a couple of specific hindrances to look out for, the primary one being your shoes.

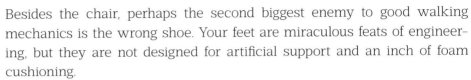

Wearing Shoes That Set You Up for Failure

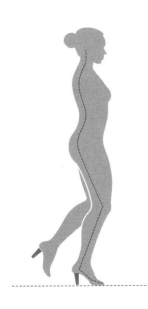

Besides the chair, perhaps the second biggest enemy to good walking mechanics is the wrong shoe. Your feet are miraculous feats of engineering, but they are not designed for artificial support and an inch of foam cushioning.

High Heels—High heels limit your ankle range of motion—similar to how a chair shortens your hips—causing your calves and heel cords to become overly short. And when you are missing normal ankle range of motion, your body compensates by turning your feet out. Do this long enough, and you've got a pretty serious negative adaptation to deal with. And it gets worse. In addition to shortening your tendons and killing your foot mobility, walking in high heels pushes your center of mass forward, forcing your spine into an overextended position. There's no getting around it: high heels destroy your gait, posture, and feet. So reserve them for special occasions if you can.

Dress Shoes—They may be shiny and make you look distinguished, but think about the block heels, firm heel caps, and stiff leather uppers. Just like high-heeled shoes, this type of spiffy footwear slowly kills the range of motion in your ankles and encases your flexible feet in shiny straitjackets. Overly stiff shoes also weaken your feet, ruin your foot proprioception (your sense of your position and movement), and screw with your walking mechanics. The raised heel tells your body that it's okay to strike with your heel out in front of you, which, upon landing, sends a shock wave through your foot and up into your skeleton. Your environment should reflect the realities of your evolved structure, and this type of shoe does not.

Heavily Cushioned Athletic Shoes—The soft cushion helps absorb some of the shock generated by heel-striking when you step too far out in front of your body, but you shouldn't be stepping far out in front of your body in the first place! And those cushy shoes are only encouraging this negative adaptation. Although the heels probably aren't as high as those of your dress shoes, they are still killing the range of motion in your feet and ankles and messing with your walking mechanics. Imagine if you encased your hands in squishy mittens and went about your day. What kind of compensations or exaggerated hand motions would you have to make to pick something up? Shoes should provide grip and protection from the elements and sharp objects, but not so much that they negatively affect your walking mechanics.

Flip-Flops—When you wear flip-flops, you have to clench your big toes to keep the shoes on your feet. This not only changes how you walk, but also causes plantar fascia strain, overly stiff ankles, Achilles problems, and various aches and pains. Look, you're not supposed to walk with your big toe and long toe pinched together. Although most sandals are flat, you're far from reaping the benefits that being barefoot delivers.

Wear shoes that affect your posture and mechanics only when the situation requires it—like at a wedding or other special event. In all other situations, choose one of these options:

Barefoot—Going barefoot as often as possible will not only help you walk as nature intended, but also increase your proprioception, improve your balance and posture, and strengthen your feet and legs. You don't have to be that creepy guy who shows up at your kid's school without shoes. (Kelly may have actually done that.) However, try to be barefoot whenever you can, especially in situations where wearing shoes is not required, like working around the house. Unless you need soles to protect the bottoms of your feet, this is a perfect time to be barefoot. We've had clients who took this advice and, in tandem with reducing optional sitting, managed to either diminish or eliminate their back, knee, and ankle problems.

Flat, Minimally Padded Shoes—Twenty years ago everything was about style, but shoe companies are starting to realize that there is a large market of people who care about their health. As a result, more flat, minimally padded (or "zero drop") shoes are coming onto the market.

While you may not consider these shoes stylish, there is always a work-around. You can wear your high heels or shiny dress shoes through the office door and out to lunch, but would anyone really notice if you changed to flat shoes at your workstation? Wearing high heels, dress shoes, or flip-flops a couple of hours a week is not going to make lasting change to your body or reverse your mechanics. Your shins might get tight and your feet might get sore, but it's nothing you can't mitigate with some soft tissue work. It's when you wear terrible shoes day in and day out that real problems set in.

The lesson here is that shoes—like other environmental loads such as sitting—affect your mechanics in unfavorable ways. So remove the casts that are weakening your feet and correct your mechanics by standing, walking, and moving with straight feet.

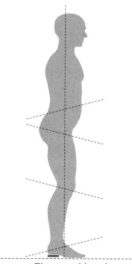
Elevated heel

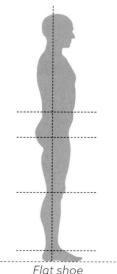
Flat shoe

Duck Walking

Big Toe Problems—Walking with your feet turned out wreaks havoc on your big toes. If you have a bony hump at the base of your big toe, known as a bunion, or you play sports or do a lot of running and have suffered from turf toe, pay close attention.

When you walk with your feet straight, your big toe bends naturally and stays in line with your first metatarsal bone, which is located right behind your big toe. This hinge is designed to permit up-and-down motion. When you walk with your feet turned out, you roll over your big toe, creating an oblique, off-axis force across your first metatarsal. To put it in simpler terms, your big toe deviates inward toward your other toes. So your big toe is not bending straight up and down; it's bending off to the side, creating a train wreck around your big toe and foot. This oblique, off-axis force is brutal and can take out even the toughest among us. Add the fact that the swollen and angry toe is about as far away from the heart as it gets and sits at the bottom of a gravity well (all your blood and other fluid tries to end up in your feet), and you can pretty much count on the swelling in that area to remain and continue causing pain until you correct the situation.

Ankle Problems—Your ankles are designed to be supported by the arches in your feet. The instant you lose that support, your ankles will search for stability by collapsing inward (pronating). This typically occurs when you stand and move with your feet turned out or if you have weak arches or stiff feet. Whether it's a mechanical or a mobility-based problem, this form of stability can cause a host of problems. Every time you take a step with your foot turned out, for example, you're essentially rolling over your ankle. This is like spraining your ankle 10,000 mini-times a day. The off-axis motion can cause Achilles dysfunction, tight calves, weakened heel cords, and an assortment of ankle range of motion restrictions.

bunion

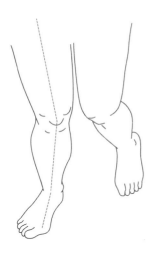

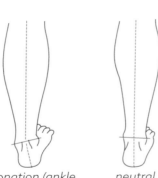

pronation (ankle collapsed inward)   *neutral*

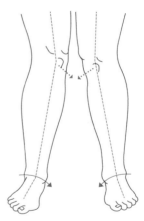

The bottom line is that your feet are the foundation for your entire body, and if you walk with your feet turned out, your arches will eventually collapse. Compromised feet set off a chain of problems that ultimately work their way up the whole kinetic chain.

Placing arch support inserts in your shoes isn't a legitimate fix because they do little to heal your weak, stretched-out, beat-down feet. The only time they can be of use is when your feet are so collapsed that you cannot walk or stand without suffering from debilitating foot pain, or if you are wearing ski boots or cycling shoes that would collapse your arches and knees when force is applied.

The good news is that your feet are built to recover and to last. While the foot is a miraculous system comprised of bony structures, connective tissues, and muscles, collapsed arches can be reclaimed. Here's your prescription:

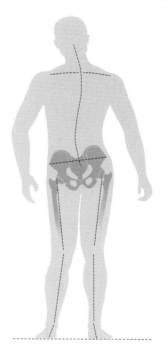

1. **Stop wearing shoes that compromise your natural walking mechanics by shortening your heel cords and stiffening your feet.** Put simply, graduate to flat shoes and try to be barefoot as often as possible.

2. **Stand, walk, and move with straight feet.** There are people out there who have structural abnormalities that prevent them from walking with neutral feet, but for the majority of us, walking with straight feet solves a lot of problems. When your feet are straight, you can walk efficiently—that is, the way your body is engineered to walk.

   As with maintaining an organized spine, walking with your feet straight takes mindfulness. To make a habit of standing, walking, and moving with straight feet, you need to constantly check your foot position throughout the day. If you catch yourself standing with your feet turned out, realign them into a straight, neutral position. If you notice that you're walking with your feet turned out, bring some consciousness to the action and start walking with your feet straight.

3. **Perform basic maintenance on your feet.** If you're missing range of motion in your anterior hips (hip flexors), lower legs, ankles, or feet, your body will naturally compensate by turning your feet out. If you don't have enough hip extension—legs moving behind the body—you will externally rotate your legs (turn your feet out) to solve the problem. The same thing happens to your ankles. When the muscles and tissues of your lower legs get stiff, this too can force your feet to turn out in compromise. To address lower leg stiffness, perform Prescription 12 on page 326. To improve hip extension, perform Prescription 9 on page 304. To improve ankle range of motion and restore suppleness to your toes and the bottoms of your feet, perform Prescription 13 on page 334.

You need your arches. So stop destroying them!

Hinging and Squatting

If we were to ask a hundred people to pick up something off the ground, about half of them would do so by bending over (hinging), and the other half would do so by squatting down. The hinging crowd would bend their knees slightly, but their torsos would come dramatically forward, in some cases coming nearly parallel to the ground. The squatting crowd would have a much deeper bend in the knees, but their torsos would remain more upright. Which method they chose would be based on a range of factors, including the weight of the object and how they were taught to pick things up as children. We would hand out gold stars not by the method they chose, because both are within our mobility wheelhouse, but rather by their technique.

Bending and squatting are natural movements that all of us should be able to perform efficiently. Assuming that everyone in both groups actually maintained the integrity of their spines while moving (no rounding, for example), the major difference between the two was simply the degree to which the hips and knees bent.

Hinging vs. Squatting
Hinging and squatting are similar movements in that you hinge from your hips while maintaining a neutral spine. The differences are the orientation of your torso and the degree of hip and knee bend. These distinctions are important because the movements, though similar in shape, have slightly different applications.

Hinging

Squatting

If you were to come to our gym and ask us to show you how to pick something up off the ground, we would teach you two exercises. The first is the deadlift, which is the gym equivalent of bending over and picking something up in everyday life. On the surface it seems like a simple movement. All you have to do is "bend" over and grab a barbell resting on the ground, straighten so that the barbell comes up to your waist, and then put it back down. If you were watching one of the world-champion powerlifters who trains at our gym perform heavy deadlifts, you would see just how efficient the human body is at bending over when done properly. Ask that powerlifter about how "simple" the movement really is, however, and she will tell you that she has been refining her hinging and lifting technique for decades.

Deadlift

Back squat

We would then pull you over to the rack section of the gym to show you the squat, which is the gym equivalent of, well, squatting in everyday life. It too seems like a relatively simple movement. All you have to do is load a barbell on your shoulders, squat down, and then return to the standing position. However, if another of our world-champion powerlifters—yes, there are a lot of them at our gym—happened to be doing squats, you would get to see someone lift the equivalent of a Volkswagen bug.

The problem for most of us is that we can't seem to connect the ideas that heavy lifting efforts in the gym and their lighter analogs in everyday life are just scaled versions of each other. In fact, you probably could get away with picking up a lightweight planter or a baby without experiencing any overt negative consequences. The caveat: it takes only one wrong movement to tweak your back. This is why you have to approach picking things up in daily life, regardless of how light they are, the same way our lifting champions pick up their 800-pound babies.

Principles of Hinging and Squatting

Principle 1: Hinge from Your Hips

Hinging the long lever of your spine from your powerful hip joints seems unnatural and foreign to most people, but this is exactly what the powerful musculature of your hips and hamstrings is designed for. It's how someone can perform an 800-pound deadlift or squat without shattering his spine or knees. When you understand how to brace your spine and harness the power of your hips, you can avoid a host of injuries that plague modern society.

Hip hinging seems like a skill that should be taught at an early age, but instead we are fed all sorts of crazy notions. A lot of people hinge from the waist (lower back) and hunch (upper back), which is a recipe for disaster. Others are taught to keep the back upright and flat, which is correct when performing a squat, but to lift with their knees rather than their hips, which is another recipe for disaster. To hinge and squat safely and effectively, you must initiate the movement by hinging from the most powerful engines in your body: your hips.

Step 1: Foot position and bracing.

To hinge correctly, the first thing you must do is organize and stabilize your spine by going through the bracing sequence (pages 82 and 83).

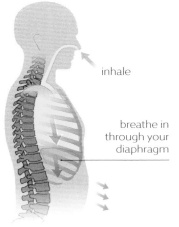
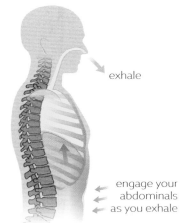

inhale

exhale

breathe in through your diaphragm

engage your abdominals as you exhale

Step 2: Shrink-wrap your spine.

Before you initiate movement, you want to increase your spinal stability by engaging the musculature of your trunk. Take a deep breath in and, as you exhale, engage your "core" by shrink-wrapping or stiffening your abs around your spine.

Step 3: Sit your hips back.

Keeping your core muscles engaged, move your hips backward as you lower your torso toward the ground. For the purposes of this drill, hinge forward only until your torso is at a 45-degree angle to the ground. This combination of movement (hips back, torso forward and down) not only allows you to keep your weight distributed over the centers of your feet, but also allows you to hinge at the hips without falling forward or backward. Your shins should remain vertical and your back flat—two important details that we will address shortly.

Sometimes it's difficult to know whether you're doing everything right, so let us point out how you should feel when you hinge forward from your hips:

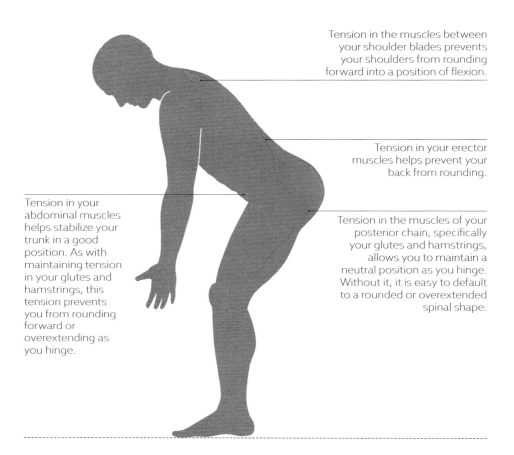

Tension in the muscles between your shoulder blades prevents your shoulders from rounding forward into a position of flexion.

Tension in your erector muscles helps prevent your back from rounding.

Tension in your abdominal muscles helps stabilize your trunk in a good position. As with maintaining tension in your glutes and hamstrings, this tension prevents you from rounding forward or overextending as you hinge.

Tension in the muscles of your posterior chain, specifically your glutes and hamstrings, allows you to maintain a neutral position as you hinge. Without it, it is easy to default to a rounded or overextended spinal shape.

Don't worry if you don't feel all this right off the bat. Remember, it takes time to retrain your body to move properly. However, there is one mistake that people make more often than most. If you don't feel the muscles in the back of your body engaging, it probably means that you didn't drive your hips far enough back before hinging forward. A good way to know for sure is to look down at your knees. If they are positioned in front of your feet, you have made this common mistake. Instead of trying to correct your position from the hinged-forward position, stand up and begin the sequence again. It's always easier to restart a movement than it is to triage a movement gone bad.

Another aspect that people often struggle with, especially those who have been deskbound for years and don't work out regularly, is keeping their trunks tight as they hinge forward. If you lack the necessary strength, as many people do, it can be beneficial to place your hands on your thighs or on a stool in front of you as you hinge. Don't worry if you lack the muscle strength needed to perform the movement properly. If you engage your core every time you go through the bracing sequence or hinge forward, you *will* get stronger and more skilled.

Although we're all wired from birth to hinge from the hips, modern society has retrained us to bend with our backs and lift with our knees, so it can take time to rewire our brains and develop the needed skills.

Principle 2: Drive Your Knees Out and Keep Your Feet Straight

To optimize your body's mechanics while hinging and squatting, it's vital to create a stable hip, knee, and ankle position. This concept shouldn't be a foreign one—in the rotation chapter, we described how to stabilize the lower half of your body by setting your pelvis in a good position: you simply squeeze your butt and screw your hips into the ground. Though it is a simple action, it sets off a cascade of vital mechanical events. In addition to putting your pelvis in a great place on your spine, it facilitates the natural outward rotation of your ankles, knees, and hips. This rotation force creates tension that stabilizes the entire lower half of your body.

You'll note that the moment you hinge forward from your hips, you lose the obvious support of your glutes. In other words, it becomes a lot harder to squeeze your butt to maintain tension and stability in your hips. Instead, you have to rely primarily on external rotation created by your smaller gluteal muscles and hip rotators to maintain stability. If you fail to do so, your knees and ankles are likely to collapse inward. To prevent this from happening, all you have to do is drive your knees slightly outward as you hinge from your hips. Just like squeezing your butt muscles and screwing your hips into the ground when standing, the cue to "drive your knees out" (or "apart") creates an external rotation force in your ankles, knees, and hips when hinging or squatting and is crucial for providing stability in the lower half of your body.

You also have to pay close attention to the relationship of your knees and ankles. Your feet should be straight and your knees should track up and down along the same vertical path, which is roughly outside of your little toes. In other words, you shouldn't see any side-to-side deviation, wobbles, or instability in your knees or ankles as you hinge from the hips and bend your knees. If they collapse inward, you're not creating enough tension (external rotation). Make sure that your big toes are on the ground so that you don't roll onto the outsides of your feet. Create just enough external rotation, or knees-out force, to prevent inward knee and ankle movement. An easy way to assess this is to watch the arches of your feet as you squat. During the squatting movement, your arches should not collapse.

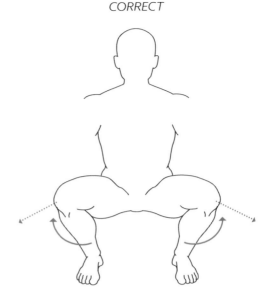

INCORRECT   CORRECT

This illustration shows what happens when you fail to create an external rotation force or "drive your knees out." Without a corresponding rotation force at the hips, your knees and ankles will collapse inward. In addition to causing all sorts of wear-and-tear on your knees and feet, this collapse creates an unstable foundation for everything above, including your hips and spine.

Creating external rotation by driving your knees out as you squat stabilizes your hips, knees, and ankles into mechanically correct positions.

Principle 3: Verticalize Your Shins

If you are able to apply the first two principles correctly, then your shins will automatically remain pretty vertical while hinging and squatting. But we've seen so many people position their shins at crazy angles when they are first learning how to hinge that we decided to make this its own principle rather than what it should be: a reminder. Whether you are bending over, squatting, or performing a combination of the two, the key is to channel your weight into the big engines of your hips and let them do all the hard work. If you allow your knees to shoot forward, well over your feet, you shift all that load onto your knees and quads. It doesn't matter whether you are squatting heavy in the gym or simply getting up off the ground: if you want to save your knees and back, you've got to employ your hips. Hip-generated motion requires your hamstrings to keep your shins from spilling out in front of you. To help ingrain this concept, let us show you some examples.

Correct way to load your hips:
Notice that Kelly's shins are pretty vertical in this photo. To achieve this while performing a simple hip hinge, all you have to do is move your butt backward (Principle 1) while driving your knees outward (Principle 2). In this position, you feel the weight of your upper body in your hips and hamstrings rather than in your knees and quads.

Loading your knees instead of your hips:
Here Kelly hasn't even hinged forward yet and things have already gone wrong. Instead of driving his hips backward to load his hips, he drove his knees forward. If he were to hinge forward from this position, his knees would be supporting all the weight of his upper body.

Loading your back instead of your hips:
While Kelly's hips are back and his shins are fairly straight in these two photos, he has lost his braced neutral spinal position. Achieving Principle 3 means little if you let Principles 1 and 2 fall by the wayside. To use your spine rather than your actual hips as a hip joint, you must either flex or overextend through your lumbar spine. This means that you are using your spine as an engine rather than as a transmission.

Lowering Your Body

Mastery of the hip hinge comes in handy for emptying the top rack of your dishwasher, but not everything you need to grab is positioned that high up off the ground. In order to truly master both the hinge and the squat (see page 111), you need to learn how to maintain all three principles while reaching all the way to the ground. This is the real challenge, because the lower you go, the more difficult it is to maintain tension in your hips and hamstrings. As you get closer to the ground, you'll probably notice that your end ranges of motion are starting to limit the quality of your mechanics and your ability to stay organized and stable.

Step 1. *Initiate a hip hinge to bend over and pick up a ball. Even if you do everything correctly—establish a braced neutral spine, drive your hips back, and keep your shins vertical, you may realize that due to years of sitting, you don't have the necessary range of motion to hinge all the way forward without compromising your form.*

Step 2. *Instead of rounding your back, drive your knees forward and come up onto the balls of your feet. This removes most of the tension from your hips. You're essentially sitting on your hamstrings, which removes some of the weight from your knees. However, if you try to stand up from this position, you will end up challenging your knee joints in a not-very-sustainable way.*

Your ankle and hip range of motion will play a bigger role as you descend deeper. If your hamstrings and hips are stiff from too much sitting, hinging over and picking something up while maintaining a neutral spine and vertical shins can seem like an impossible feat. To reach the ground, you may feel that you have to compromise one or the other. We prefer that you always prioritize your spine. However, there is a right way and a wrong way to cheat.

Step 3. *Instead of standing straight up from your "cheat" position, drive your hips backward by driving your knees outward and getting your shins vertical. Notice how Kelly's back is flat and protected.*

Step 4. *Having established a braced, stable position, you can now stand up without putting unneeded strain on your back or your knees.*

Step 5. *Return to the upright position and reestablish your base standing position.*

You can even apply this same strategy to getting up off the ground.

And if you're kneeling on the ground, which is similar to the squat shape, you can use the same strategy—drive your hips back, get your lead shin as vertical as possible, and then stand up. Here are a couple of options:

OPTION 1

OPTION 2

Hinging Mechanics

Now that you understand how to load your hips and hamstrings and use your knees to adjust for position, it's time to focus on perfecting your hinging technique. It's important to note that, due to severe tissue restrictions, only a fraction of the people who have subjected their bodies to the deskbound life will be able to perform this movement correctly right out of the gate.

Step 1. To begin, go through the bracing sequence (pages 82 and 83). Position your feet underneath your hips, just inside your shoulders' width. The toes of both feet should be pointing straight ahead. With your feet neutral, your abs braced, and your shoulders and hips stabilized, you're set up for a safe and healthy hinge. Notice that the ball Kelly is about to pick up is close to his body, between his feet. Anytime you want to pick something up, try to position the object as close to your feet as possible, which puts the object underneath your hips. The closer it is to your body, the more effective your mechanics will be.

Step 2. Keeping your shins as vertical as possible and maintaining a neutral spine, sit your hips and hamstrings back and tilt your torso forward. To maintain a stable ankle, knee, and hip position, create a small external rotation force by screwing your hips into the ground through your feet.

Step 3. With your weight distributed over the centers of your feet, keep a slight bend in your knees—still driving them out slightly—and continue to drive your hips back. As you lower your elevation, allow your torso to come forward slightly so that you can reach the object you're lifting. Notice that Kelly's back is flat and his feet are still flat on the floor. If you lack range of motion, keeping your feet flat may be difficult to manage. Do not compensate by rounding your back.

If you fall into this category, you can "cheat" by utilizing the lowering technique shown on pages 102 and 103, but as you focus on your posture and improve your mechanics by using the techniques presented in this book, it is important to know what to strive for. Clean hinging mechanics are a hallmark of human movement and an excellent predictor of hip and lumbar function.

Step 4. *Before you attempt to lift the object, verticalize your shins (Principle 3). You do so by simultaneously raising your hips and driving your knees back. This creates tension in your hips and hamstrings and ensures that your knees and back aren't the primary lifting engines.*

Step 5. *Maintaining a neutral spine, extend your hips and knees. Think about squeezing your glutes and keeping tension in your hips as you stand upright.*

Step 6. *As you stand upright, squeeze your glutes and reestablish your base standing position.*

Hinging Faults and Corrections

There's no getting around it: bending claims more backs than any other movement. If you have a habit of hinging at the waist and bending over with a rounded back—whether you're tying your shoes, bending over a workbench, or picking up a pillow—you're susceptible to back tweaks and possible low back pain.

Here's a common scenario: After sitting all day at work, your body is tight and your mind is on autopilot. Your desk is a mess. As you're cleaning up, you accidentally knock a pile of papers onto the floor. You hastily bend over to pick them up, and on the way up you feel something go horribly wrong. You let out a gasp and immediately reach for your lower back. You just suffered a back tweak.

To clarify, a back tweak isn't fun, but it probably isn't a serious injury. You didn't automatically herniate a disc or break your back. In all likelihood, you just experienced a soft tissue strain. But it probably feels like a demon took up residence in your back because it hurts like hell. Your back basically goes into shutdown mode, and the surrounding muscles go into spasm, causing discomfort and pain.

You already know how to avoid this scenario: be conscious of your position and bend over by hinging from your hips while keeping your spine organized and stable. It's that simple. But we know that some of you will default to old movement patterns and commit fundamental errors. You're human, after all. So let's turn our attention to the two most common hinging faults—the waist-hinge fault and the too-much-extension fault. We will discuss why these faults occur and, most important, outline strategies for correcting the errors.

The Waist-Hinge Fault

By far the most common mistake that people make when bending over is hinging from the waist rather than the hips. Notice in the photo below that this not only puts a bend in the lower back, but also causes the upper back to hunch. So many people utilize this improper technique because it is simply the path of least resistance. You essentially bend over with no tension and allow your upper body to collapse over your lower body to the limits of your soft tissues. When you do this, you're asking your back to perform the job of your hips—something it was not designed to do. Take a second to compare the size of your spinal vertebrae to the size of your pelvis-femur system. Which of these do you think is more capable of hinging under a 1,000-pound load?

Mechanical Fix

If you are unable to perform a correct hinge due to improper technique, master the bracing sequence (pages 82 and 83), learn how to load your hips and hamstrings (page 96), and practice hip hinging mechanics (pages 106 and 107).

Back Tweak Fix

The goal is to learn how to hinge properly and avoid back tweaks, but if you tweak your back and are in pain, use Prescription 5 on page 278 to put yourself on the road to recovery.

Range of Motion Fix

If you are unable to perform a proper hinge due to soft tissue restriction or tight hamstrings, use your knees to adjust for position (pages 108 and 109) and perform Prescription 5 on page 278 to improve your soft tissue mechanics.

The Too-Much-Extension Fault

The second most common mistake that people make when bending over is to overextend their backs. Even if you have every intention of keeping your back flat, if you fail to maintain trunk tension and set your pelvis in a neutral position, overextension is the likely result. The moment you sit your hips back, your pelvis tilts forward and you lock in a bone-on-bone position of stability. And if you sit in your chair with this spinal shape as well, you have pretty much set yourself up to reproduce this pattern of hinging.

Hinging with an overextended spine works for a while, and you'll fool your friends into thinking that you aren't rounding, but it's not quite what the spine evolved to do. It is a little-known fact that most of the spine tweaks that occur in sports, at the gym, and in yoga class are actually overextension in nature. Slamming the doorstops of your spine together in extension is bad enough while you're sitting still, but adding load and speed to this shape in the form of a dynamic hip hinge isn't sustainable over the long haul. One reason this spinal pattern is so common in training movements is that our sitting-shortened soft tissues drag our spines into this shape when we try to stand up. It's like a loaded spring chronically pulling your spine toward too much extension.

In most cases, this error occurs as you initiate the hinge. So, if you unlock your pelvis at the start of the movement and feel your back muscles tighten up and your pelvis roll forward, return to the standing position and get reorganized. This is how you break the cycle of poor movement patterns: if you catch yourself in a bad position during a movement, don't continue moving poorly. Stop, reset, and then try again. After a while, you won't have to stop and reset your position. Your training will kick in, and moving correctly will become as easy as pie.

Mechanical Fix

Although tight hip flexors can predispose you to overextending, this fault is more technique than anything. You need to understand the bracing sequence (pages 82 and 83) and ramp up your trunk tension as you load your hips and hamstrings. Repeat Steps 1 and 2 over and over until you can do them without overextending. It might take you a few tries, but you will get it.

Overextended-Back Spasm Fix

In order to get back to baseline, you need to break those spasms—see Prescription 5 on page 278—and address your faulty mechanics.

Squatting Mechanics

As with walking and hinging, squatting is a movement that you perform an incredible number of times a day. You squat every time you get up out of bed, use the toilet, sit down to a meal, work from a chair, drive a car, or sit on the couch. In a nutshell, you squat every time you lower and raise your center of gravity with an upright torso, which encompasses a lot of life and sport activities. This makes the squat one of the most universal and transferable human movements there is. If you understand how to squat with good form, you can apply the technique to anything that embodies the squat shape: sitting, jumping and landing, picking up heavy objects—the list goes on and on. In short, learning how to squat with good technique helps you acquire new skills that involve squatlike movements, of which there are a lot.

It bears repeating that squatting shares all the same movement principles as hinging: you maintain a neutral spine, hinge from your hips, and then use your knees to adjust for depth. The only things that separate the two movements are the angle of your torso and the degree of hip and knee flexion. When squatting, you typically need to move with a more upright torso. When hinging, the goal is to keep your hips high and your torso more parallel to the ground. Sometimes you'll hinge over to tie your shoes, and other times you will squat or kneel to the ground to tie them. And other times you might combine the two movements, hinging and then dropping into a squat.

You already know how to hip hinge without compromising your spinal mechanics, so let's see how the principle of creating tension in your hips applies to squatting. We'll start with the easiest and most common type: the chair squat, also known as the box squat. To perform this sequence, you will need a chair or bench with a hard surface and without wheels or casters, or a sturdy box that is at least 18 inches high.

Many of the root causes of common musculoskeletal problems stem from performing movements like squatting to a chair without giving any thought to the underlying mechanics. Loose lug nuts on your car tires aren't really a problem for short distances at slow speeds, but given enough distance and velocity, you can be sure that the wheels will fall off. We feel the same way about squatting with a disorganized spine and with no attention paid to what your knees and hips are supposed to be doing.

Chair Squat (Box Squat): Sitting Down

Step 1. *Position your feet just outside your shoulders' width, with your toes pointing straight ahead. The front legs of the chair should be about 6 inches from your heels. Now go through the bracing sequence (pages 82 and 83).*

Step 2. *After you've organized your spine, the next step is to load your hips and hamstrings. You do so by driving your hips and hamstrings back, bending your knees slightly, and tilting your whole torso forward. Imagine that you are trying to shut a car door with your bum because your hands are full. As you do this, drive your knees out, keeping your shins as vertical as possible. The moment you lean forward and shoot your hips back, you need to ramp up your trunk tension to keep your spine stabilized and neutral. Keep your gaze focused about 5 feet in front of you and your weight distributed over the centers of your feet.*

Chair Squat: Getting Up

Step 5. *To complete the sequence, simply reverse the steps. From the seated position, allow your torso to come forward slightly until you feel tension in your hips and hamstrings. Maintain pretty vertical shins. Don't trade forward-tilting shins for unloaded hamstrings.*

Step 4. *From here, you can transition to an upright position. This is the ideal seated posture.*

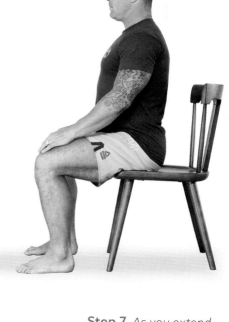

Step 3. *Lower your hips and bend your knees—still keeping your shins verticalized. Without looking, try to position your butt on the front half of the chair. Try to sit on the bony parts of your pelvis (your ischial tuberosities).*

Step 6. *Load your hips and hamstrings by lifting your butt off the chair. Mid-squat, it should be impossible to tell whether you are going up or down; the sequence is the same. As a "new" squatter, your torso may incline more forward than you might expect.*

Step 7. *As you extend your hips and knees to stand upright, bring your shins fully upright, squeeze your glutes, and reestablish a braced spine. You've just exposed your spine to significant forces. Reorganizing your spine at the top of the movement will set you up to transition well into your next shape, like walking.*

Deep Squat

The chair squat is what you might consider a modern-day squat because the chair limits your range of motion. Squatting well to this height will be a staple in your movement pantry. But if you have to squat lower or get up off the floor, both of which our bodies are designed to do, you have to learn how to execute a deep squat. It starts out just like the chair squat, but instead of stopping your descent a couple of feet off the ground, you keep going so that your hamstrings are flush with your calves and your hips drop below your knee crease. If you were sitting by a campfire and the ground was wet, the deep squat would be your position.

It is important not to compromise your form in order to achieve a full squat. You want your heels on the ground, your toes pointing forward, your knees out to the sides, and your back flat. If you can't achieve this position without compromising your form, don't worry; most deskbound warriors are in the same boat. On the next page we will show you how to work toward improving your squat depth. And the simple steps of sitting less and employing daily mobility work will help you achieve the ultimate goal of reclaiming your birthright of squatting to full depth.

Supported Squat

If you're unable to achieve a full-depth squat due to range of motion restrictions, spend 10 minutes hanging out in the supported squat positions shown opposite. We call this the 10-minute squat test. For the best results, you want to remain in the squat for the entire 10 minutes. You can shift from one supported squat position to another or stand up momentarily to stretch your legs, but the majority of those 10 minutes should be spent in the bottom of the squat. When you are able to sit in the bottom position without holding onto anything or complaining about your burning shins, you've passed the 10-minute squat test!

If your range of motion is severely limited, this exercise can be quite challenging. If necessary, you can break up the 10 minutes into several sessions over the course of a day. But remember, the key to improving your squat depth is to spend time in this shape. Spending 10 minutes a day in a deep squat is a great life goal.

Range of Motion Fix
In addition to hanging out in a deep squat, you can use Prescription 9 on page 304 and Prescription 10 on page 312 to improve your squatting range of motion.

Squatting Faults and Corrections

Most people don't think about their form when they sit down or get up out of a chair, but the same ugly movement patterns that cause so much trouble when standing, walking, and bending also frequently plague the squat. These faults include turning your feet outward and allowing your knees and ankles to collapse. When you stop and think about how many times you sit down and stand up over the course of a day, you can imagine the cumulative effects.

To make matters worse, with each improper squat that you perform at your desk, you're ingraining a poor movement pattern that will find its way into more advanced squatlike movements. If you squat poorly to a chair, we don't even need to see you lift a heavy weight at the gym to know that you will be compromised. The way you do one thing is the way you do everything. If you want a body that's rock-solid under any kind of load, then moving well under no load is the path to that goal.

While utilizing the proper squatting technique described previously will set you down the right path, it is impossible to focus on your form all the time, especially when you're engrossed in work. To help keep you from reverting to old patterns after you put down this book and move on to other aspects of your life, in this section we highlight the most common squatting-shaped faults and how to correct them. Studying these faults will dramatically increase your chances of spotting your own movement errors as they occur. And, since the ultimate responsibility for teaching your kids to move resides with you, being able to spot common movement faults in your kids is a double win.

Knees-Forward Overextension Fault

When it comes to squatting, the most common mistake that people make is to shoot their knees forward to start the movement. This crazy knee motion leads directly to an overextended spine that too closely resembles exotic dancing. You've certainly seen this fault at a restaurant or around the office when someone gets up out of a chair. It looks like this:

You scoot your butt forward to the edge of your seat, which places your knees well in front of your shins. Then you stand up, hearing all kinds of creaks and cracks in your knees. Because your knees are mechanically compromised in this tenuous position, your lumbar spine tends to buckle forward. By the way, this "artificial" movement pattern exists only as a function of our chair-based environment. Add high heels and a pencil skirt to this scenario, and it's no wonder the United States leads the world in back surgeries and knee replacements.

As you know by now, you want to load your hips and keep your shins as vertical as possible. The flawed technique just described places all of your weight on your knees when they are in a poor mechanical position. While it probably won't cause injury in the short term, constantly loading your knees in this manner burns through your duty cycles and can cause knee pain and even degeneration in the long run. And don't get us started on the dysfunctional movement pattern that you're ingraining. As Juliet likes to say to our children when they are expressing behavior of dubious consequence, "Make a better decision."

Mechanical Fix

Go through the bracing sequence (pages 82 and 83) and get your belly tight as you load your hips and hamstrings. To keep your knees from moving forward, get your shins as vertical as possible and drive your knees out slightly as you initiate the movement. Think about pulling the backs of your knees toward your body as you start to stand up.

INCORRECT   *CORRECT*

Knee and Ankle Collapse

As you know, there are several steps to performing a proper squat. The second most often forgotten step, trailing right behind keeping your shins vertical, is creating an external rotation force from your hips by driving your knees out. Whether you are squatting to the ground or standing up out of a chair, when you forget to create this stability, your knees and ankles tend to collapse inward, a fault commonly referred to as "valgus knee collapse."

It is important to note that this fault often goes hand in hand with not having straight feet when you move. If your toes are pointing outward when you are preparing to squat, as shown below, your ankles will tend to collapse inward, flattening the beautiful arches of your feet. And when your ankles and arches collapse, your knees follow suit.

Mechanical Fix
Position your feet straight and create an external rotation force by driving your knees out slightly.

CORRECT

INCORRECT

"Butt Wink" Fault or Wobbly Low Back

Another common squatting fault is the "butt wink," where your pelvis tucks underneath your body as you lower into the bottom of the squat. "Butt wink" is coaching shorthand for the loss of a stable pelvis-lumbar relationship, and it causes multiple small train wrecks in and around your lower back. When you commit the butt wink fault during a weighted squat, you are systematically loading structures of your body that aren't designed to be both columns and wheels at the same time.

INCORRECT

Step 1. *You set up to perform a squat by driving your hips back, but you fail to brace your spine by using the bracing sequence. As a result, your spine overextends and your pelvis tilts forward.*

Step 2. *Because you failed to complete the bracing sequence, your entire body lacks stability. As you lower into the bottom of the squat, your pelvis tilts back and tucks underneath your body, like a dog with its tail tucked between its legs.*

Step 3. *With your body still lacking stability as you stand back up, your pelvis tilts forward again and you return to your overextended spinal position.*

Mechanical Fix

Initiate the squat by loading your hips and hamstrings while maintaining a braced neutral spine. Think about driving your knees out slightly as you lower your elevation. If you're limited by your mobility, don't sacrifice your form to reach the bottom of the squat. Instead, reduce the depth of your squat or compromise your knees' leverage by allowing them to translate forward so that you can achieve your desired depth.

Range of Motion Fix

If you're butt winking due to a range of motion restriction, you'll need to spend some time working on your squatting range of motion. Use the supported squats demonstrated on pages 114 and 115 and perform these mobility prescriptions to improve your range of motion: Prescription 9 on page 304, Prescription 10 on page 312, and Prescription 13 on page 334.

Sitting at the very bottom of the squat with your pelvis tucked underneath your body is not necessarily bad for you. In fact, this is how we are supposed to squat when using the toilet or just working on the ground, like cooking around a fire. It's when you add movement (rotation of your pelvis), load, speed, or repetitions to the equation that tucking becomes a problem. The natural reversal in your lumbar spine that comes from sitting on your haunches is actually quite restorative to a spine that is habitually biased into extension by our sitting environment.

When you perform the 10-minute squat test (pages 114 and 115), don't worry about being in an "active" spinal shape. You should have sufficient range of motion in your hips that you can reclaim a braced spine even after relaxing at the bottom of the squat. Imagine that you were squatting all the way down and someone handed you something heavy. Would your spine pass muster? As you transition from safe to full capacity, you ultimately need to be working toward the ability to squat all the way down and generate a braced spine. After all, any two-year-old child can do it.

One-Hundred-Year Shoulders

If you've read and practiced the bracing sequence on pages 82 and 83, you already know how to organize your shoulders while standing. If you're carrying something heavy, you want to create a lot of rotational stability in your shoulders, and if you're carrying something light, you need only a little rotational tension. But the concept should always be there. It's like a suit of armor protecting your joints and tissues from abuse.

Though this concept is relatively simple, it can be difficult to apply, especially when you're living the deskbound lifestyle. Remember, your spine is the chassis for your hips and shoulders. Put simply, when your spine is out of alignment, it is nearly impossible for the engines of your hips and shoulders to function well. For example, if you're in a position of global flexion, chances are your shoulders will destabilize and default to rolling forward.

In order to break this cycle, let's focus first on establishing the proper shoulder position for everyday activities, such as texting and carrying a briefcase, and then progress to finding shoulder stability in more taxing movements.

CARRYING

Stable Shoulder Shapes

When it comes to organizing and stabilizing your shoulders for most arm movements, there are three basic positions to consider: arms down, arms horizontal, and arms up overhead. Whether you're going about your day or lifting weights in the gym, chances are you'll be expressing movement from one or more of these three shoulder positions. If you understand how to get organized in these shapes, you have the building blocks to create safe, stable positions for most shoulder movements. Picking a kid up off the ground? No problem. Pushing a heavy cart down an aisle? No problem. Lifting your suitcase up to the overhead bin in an airplane? You guessed it, no problem.

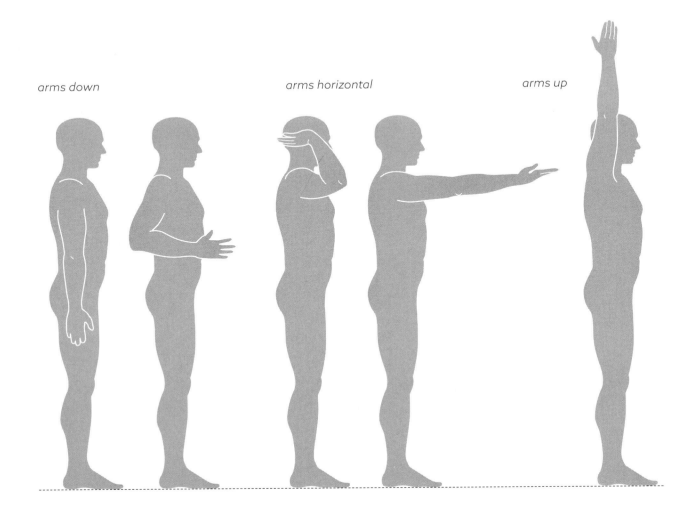

arms down

arms horizontal

arms up

Arms Down by Your Sides

If you've already mastered the bracing sequence, it will be easy to learn how to organize your shoulders and carry something either at your side or in front of your body. To begin, go through the bracing sequence, paying special attention to Step 3, where you turn your palms toward the sky. Once you feel the tension in your shoulders, lower your arms to your sides and keep your thumbs pointing forward. From this position, try picking up something light, such as an empty laptop bag, with one hand. First, notice the tension in your shoulders. Because the object is light, the tension you feel should be similar to the tension you felt right after completing the bracing sequence. If that tension disappeared, put the object down and start over. Next, check to make sure that your thumb is pointing straight ahead, your elbow hasn't flared out away from your body, and your shoulders haven't rounded forward.

If all is well, try to create the same stable shoulder position with your arm bent. For example, grip a laptop in one hand with your palm facing upward and your thumb turned out to the side. As you know from the bracing sequence, this hand position—thumb out and palm up or forward—creates external rotation, which increases the stability and tension in your shoulder. Again, take note of this tension and check your posture.

Finally, let's move on to a two-handed carry. We use a chair in this example, but if a chair proves to be too heavy for you, you could hold an object such as a laptop with both hands out in front of you.

The mechanically simplest way to carry an object is to turn your palms up. The palms-up carry naturally orients your shoulders into a more stable shape. It isn't always possible, however, if you're carrying a chair, a box, or something with handles, like a laundry basket. In these situations, you can still add tension to your shoulders by keeping your elbows in tight to your body and externally rotating your shoulders through your hands into the handles or object you're carrying, just as you would when turning your palms toward the sky. Your hands won't move because they're fixed, but the motion will generate the required tension. If you think that this sounds like screwing your hips into the ground through your feet, you've got the concept.

Arms bent, hands out in front of you, is also the position from which you type. Typing with little stability in your shoulders sets up the entire kinetic chain of your arm for potential dysfunction. To avoid this, always create stability in your shoulders before pecking away at the keyboard: turn your palms upward to create external rotation and tension (are you starting to make the connection?) and then maintain that same stable shoulder position as you flip your hands over to type.

No matter what you want to carry or how you position your hands, the key is to add tension to your shoulders by generating an external rotation force before picking up the object or setting up. If you feel your shoulders rounding forward as you're working at your desk, don't stay there. Take a moment and reset your shoulders into a good position.

The Right Way to Carry a Bag

You probably never think about how you hold your bag, but you should. Carrying a bag the wrong way can corrupt your posture and cause pain and restrictions. For example, let's say you carry a laptop bag with the strap draped over one shoulder. With all the weight resting on one side of your body, you compensate by lifting your shoulder, which throws off your spinal alignment. Even worse, you have to roll your shoulder forward to keep the strap from slipping off. Carrying a handbag in the crook of the arm is another common technique that can cause similar disruption to the shoulder and spine. And the crook of your arm isn't very good at bearing the weight of a heavy bag. Look at anyone carrying a purse with her elbow, and we guarantee that you will see a shoulder positioned well out in front of the body.

Fortunately, these common errors are easy to avoid. All you have to do is use the straps as they were intended. For example, if you're carrying a backpack, use both straps instead of just one. If you're carrying a laptop bag or handbag, sling the bag diagonally across your body. It's called a cross-body bag because it's designed to be carried across the body! This takes the load off your shoulder and lower back, allowing your trunk muscles to do the work. And instead of carrying your purse in the crook of your arm, try gripping it by the handle. You might also try switching the position of the bag or your grip often and adjusting the straps to minimize swing.

INCORRECT

CORRECT

Arms Horizontal

How much time do you think you spend on your smartphone? The data is somewhat convoluted, but according to recent statistics, U.S. smartphone users spend an average of two hours per day surfing the Internet, social networking, gaming, texting, and taking and receiving calls.[1] Let's err on the conservative side and assume that the average person spends 90 minutes a day on the phone. To put it in perspective, that's 23 days a year and 3.9 years of the average person's lifetime spent using a phone.

Now picture yourself using your phone. It's difficult to do without looking down, craning your neck, and rounding your shoulders. Left unchecked, that's nearly *four years* of your life spent in this compromised shape, and that doesn't include working on a computer, watching TV, eating, and so on.

But here's the good news: there is a simple way to hold your phone that will not compromise your posture. As you would when setting up for any other movement or position, go through the bracing sequence and set your shoulders in a neutral position. Then, keeping your shoulders organized and your head in line with the rest of your body, bring your hands in front of your face. Again, your thumbs are pointing outward and your elbows are in tight to your body. In this situation, the goal is to get your biceps to face the ceiling. This is how your body's mechanics would suggest that you use the phone.

If you're talking on the phone, same thing: try to position your biceps toward the sky and keep your elbow in tight to your body. It's that simple.

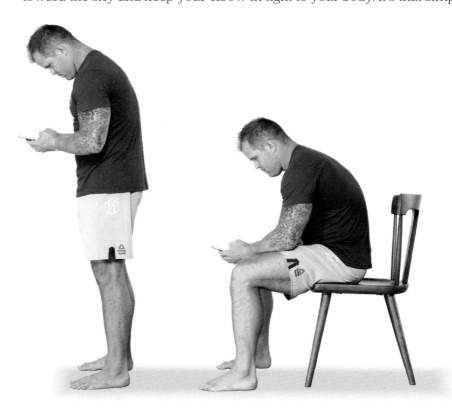

Although we are using a phone as an example to describe how to get organized in a neutral, stable position when your arms are horizontally oriented, it's important to realize that this is how you should organize your shoulders whenever your arms are extended out in front of you. For example, say you're holding a kid with a poopy diaper or lifting a weight. You still want to stabilize your shoulders by keeping your biceps up and your elbows in tight to your body.

Arms Up

When people perform overhead movements, a lot of them pay close attention to their spinal position and ignore the position of their shoulders. This is a big mistake, and more obvious if you are lifting something heavy, but insidious if you are performing repetitive "light" overhead movements like swimming. Luckily, finding shoulder stability in the overhead position is just as easy as it is in any other position. The tricky part is getting your hand positioning correct.

When your arms are out in front of you or down at your sides, it is easy to remember the cue "palms to the sky." To avoid instability when your arms are overhead, remember the cue "palms facing each other" or "thumbs pointing behind me." Sometimes it's easier to think about rotating your armpits forward to create a solid overhead position.

PUSHING AND PULLING

Whether you are pushing a car to a gas station or pulling yourself over a ledge, creating stability in your shoulders through external rotation is mandatory. What complicates matters slightly is not your pushing or pulling action, but rather the object you are attempting to push or pull. Here are the three scenarios that we're going to cover:

1. Pushing or pulling something that you can grip with both hands, such as a barbell or shopping cart

2. Pushing something that you can't grip, such as the ground or the bumper of a car

3. Pushing or pulling something that you can only grip with one hand, such as a dumbbell or duffle bag

"Break the Bar": Pushing or Pulling Something That You Can Grip with Both Hands

If we were to ask a gym rat to name something we commonly push and pull by gripping it with both hands, she would most likely say a barbell. Someone who doesn't go to the gym might say a shopping cart or a lawnmower. Fortunately, teaching someone how to push and pull properly requires only one explanation. As a matter of fact, you can apply this technique anytime you push or pull something that you can grip with both hands. The sole caveat is that the object must be sturdy enough not to bend or break under the pressure.

Just as with carrying, the goal is to generate an external rotation force to provide tension and stability to your shoulder complex. You learned in the bracing sequence and again in the carrying section that you can accomplish this by extending your arms out to your sides and turning your palms toward the sky. Well, when you're gripping something with both hands, such as a barbell, this isn't possible. However, nothing is stopping you from *trying* to turn your palms to the sky. While the bar doesn't actually bend or break, you'll find that the "breaking the bar" technique still brings your elbows in tight to your body and generates tension in your shoulders. After practicing it for a while, you'll discover that, unlike turning your palms upward, you're in complete control of how much tension you generate.

This is the beauty of gripping a solid object with both hands: it's a closed-loop system. "Breaking the bar" is a good name to use when someone is about to bench-press 500 pounds because he needs to generate a whole lot of external rotation to keep his shoulders stable under the massive load. Before executing the lift, he wants to try to break the bar in half. However, the name doesn't fit so well when someone is pushing or pulling an empty shopping cart. Remember, the amount of external rotation you generate should be in line with the amount of weight you find yourself pushing or pulling. If you're pushing an empty shopping cart, you need very little external rotation. As the shopping cart fills up, you increase the "bend."

People who teach workplace safety generally consider generating stable shoulders for pushing to be too hard to teach. They usually encourage workers to pull large loads instead. Why? One of the most common mechanisms for shoulder injuries in the workplace is pushing a heavy load. We are pretty confident that if you can organize your spine and shoulders before you try to push a wall down, very little of your person will be harmed, except your ego.

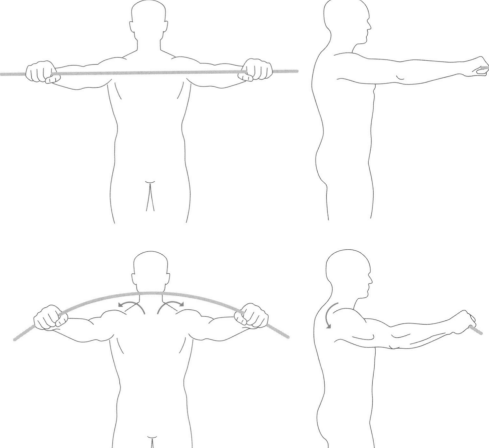

For this exercise, you'll need a PVC pipe or a wooden dowel.

Step 1: *Form your grip on the pipe at about shoulder width. You can hold it out in front of you, down by your hips, or over your head. The technique is exactly the same.*

Step 2: *Twist your hands into the pipe as if you were trying to break it in half. Be sure to initiate the rotation from your shoulders. Unless you're bench-pressing or doing pull-ups, you need to cultivate only a slight external rotation force for most everyday activities.*

"Screw Your Hands into the Ground": Pushing Something That You Can't Grip

Ask someone to name a surface that she can't grip but frequently pushes off of, and she will probably say the ground. This is called a push-up, so let's use it as an example. If you ask that same person to show you a push-up, she will likely drop to the ground, position her hands roughly shoulder width apart, and begin cranking out repetitions. Unfortunately, you may see her elbows flare out to the sides and her shoulders round forward. Translation: she is working hard without understanding some basic body principles. The repercussions may not be immediate, but they will certainly be felt in the long term.

When we think of a push-up as a way of getting up off the ground, we can see that this kind of pushing isn't just about getting buff; it's a lifelong skill. As we've said, the number-one reason people end up in nursing homes is that they can no longer get up off the ground independently. Even yogis knew that this skill was worth practicing. It's no accident that the technique behind the signature yoga move Chaturanga looks an awful lot like a good push-up.

To turn an ugly pushing situation into a healthy one, all you have to do is add external rotation to the equation. Since you can't turn your palms toward the sky as you do in the bracing sequence or "break the bar" as you do when pushing something that you can grip, this situation might seem tricky, but it's not.

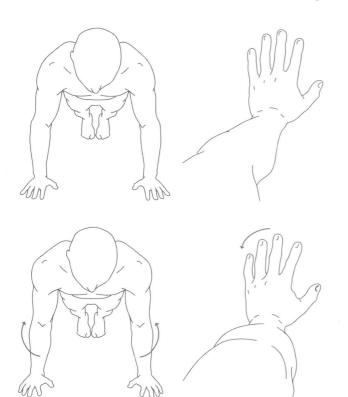

Step 1: *Get into the push-up position with your hands about shoulder width apart and your fingers pointing straight ahead. If you can't support your weight in this position, place your knees on the ground. If this is still too difficult, stand up and place your hands against a wall.*

Step 2: *Externally rotate your shoulders so that your biceps are pointing straight ahead. An easy way to accomplish this is to screw your hands into the ground. Your hands remain stationary, but you apply pressure as though you were trying to screw them into the ground through your shoulders. The trick is to apply enough external rotation force that your elbows move closer to your body and your shoulders stabilize, but not so much that your hands start spinning like the hands on a clock. Not everything that you push, pull, or carry will be as heavy as your torso, so it's important to pay close attention to the tension you've just created. Feel how tight your arms are to your body and the newfound stability in your shoulders.*

Pushing or Pulling Something That You Can Only Grip with One Hand

Let us clarify what we're talking about here. We don't mean establishing a one-handed grip on something stationary, such as a pull-up bar. If the object is fixed and you choose to establish a one-handed grip, you can use the previous technique for generating an external rotation force and creating stability in your shoulder. What we're talking about is gripping something that is movable, like a dumbbell or briefcase. In this scenario, the "break the bar" technique won't work because the object would just spin in your hand. "Screwing" your hand into the object won't work, either, for similar reasons.

To generate external rotation in this situation, you need to revisit the bracing sequence (pages 82 and 83). Because the object won't allow you to generate the rotational boost you'd get from grabbing a fixed object, you must set your shoulder prior to gripping the object. This requires you to go through the bracing sequence and set your tension level ahead of time. For example, if you're going to lift a light dumbbell, you might need only slightly more tension than you would when preparing to work at your standing desk. However, if you're going to push or pull a heavy dumbbell, you'll want to take up all the slack in your shoulder tissue first.

A great real-world example is lifting a jug of milk out of the fridge. Have you ever noticed that once the milk comes off the shelf, it drops a few inches before your "muscles" kick in? This isn't because the milk jug is heavy. The drop occurs largely because the falling jug has to take up the slack in your poorly organized shoulder. Talk to anyone who has a shoulder injury and he will tell you that screwing his shoulder into place prior to lifting that million-pound carton of milk is the only way he can do it pain-free.

Shoulder Faults and Corrections

Because we work and live in the physical therapy and strength and conditioning worlds, we've seen more shoulder injuries than we care to count: impingements, dislocations, rotator cuff tears, labral tears, and biceps tendinitis, to mention a few. These injuries are painful, they take forever to heal, and, because we use our arms and hands for pretty much everything, they are a major bummer. When you're faced with one of these injuries, even everyday activities like brushing your teeth and driving become difficult and sometimes impossible to do. The good news is that, like the rest of your body, your shoulders are built to last a hundred years. When a deskbound athlete comes to us asking for advice about an ache, pain, or nagging shoulder injury or asking why his shoulder hurts every time he moves his arm, we always say, "First things first, we have to get you out of that internally rotated shoulder position."

Internal Rotation Fault

The internal rotation fault occurs when your shoulders roll forward into a slouched position. This can be due to poor habits or poor shoulder range of motion. It largely stems from spending too much time in a slouched-over, rounded-back position.

This fault can cause two things to happen:

1. **Overstretched and adaptively stiff shoulders:** Imagine sitting with slouched posture, your back hunched and your shoulders rolled forward. If you take a moment to think about it, you'll notice the weight of your arms pulling the heads of your humerus bones (upper arm bones) forward and out of your shoulder sockets. In addition to putting your rotator cuff musculature in a tough working position, this shape relies mostly on the shoulders' passive connective tissues to maintain the joints' integrity. It's like pulling on both ends of a T-shirt; after a few minutes, the shirt stretches out and loses its shape. Well, this is exactly what happens to some of the connective tissues of your shoulders: over time they can become lax. What's more, the fronts of your shoulders and chest muscles adapt to this position by becoming stiff, essentially forming a cast around your poor posture choices.

   If you have loosey-goosey shoulder capsules and adaptively stiff chest and shoulder muscles, use the mobility prescriptions listed on the opposite page and fight like hell to keep your shoulders stable. Your tissues will come around eventually.

2. **Missing shoulder range of motion:** If you're missing shoulder internal rotation range of motion—think moving your arms behind your back—the heads of your humerus bones will translate forward in the joint sockets and your scapulae (shoulder blades) will wing over to compensate and allow you to continue to perform crucial tasks. In other words, missing internal rotation range of motion will cause your whole shoulder complex to collapse forward internally. To help you understand how this fault can occur, we'll use a simple example.

   Notice that Kelly is lying on his back and dropping his palm to the ground. This is internal rotation of the shoulder.

If you move beyond the limits of your mechanics and keep internally rotating your arm, your shoulder will roll forward. So, if you're missing shoulder internal rotation, you will compensate by rolling your shoulder forward.

This is a brilliant adaptation because it gives you more range of motion to move your arms. The structure of the shoulder is set up to maintain vital movements like feeding yourself. Having a scapula is like having a whole other shoulder joint behind the one you think of as your shoulder. Can you imagine if your shoulders weren't like this? If your shoulders became as stiff as the typical deskbound person's hips, you would not be able to function. The problem is that this backup movement generator allows you to get away with poor movement, with few mechanical consequences other than lost shoulder power, until your tissues finally throw in the towel. This is why physical therapists and chiropractors treat the "shoulder joint" as a system that includes the scapula.

An internally rotated shoulder is a weak and vulnerable shoulder. So, if tissue restriction is the limiting factor, get to work on improving this internal rotation range of motion like it's your job.

Mechanical Fix:
Organize your shoulders in a neutral position and create an external rotation force to stabilize the position. For most shoulder movements, keeping your elbows in tight to your body and pointing or rotating your thumbs away from your body will stabilize your shoulders in a good position.

Range of Motion Fix:
Use Prescription 4 on page 272 to wake up the long-locked tissues around the backs of your shoulders and restore suppleness and range of motion to your chest and the fronts of your shoulders. To improve shoulder internal rotation, perform Prescription 3 on page 266.

Section 4

Standing Workstation Guidelines

The Active Workstation: Creating a Movement-Rich Environment

From Sitting to Standing: How to Transition Safely to a Standing Workstation

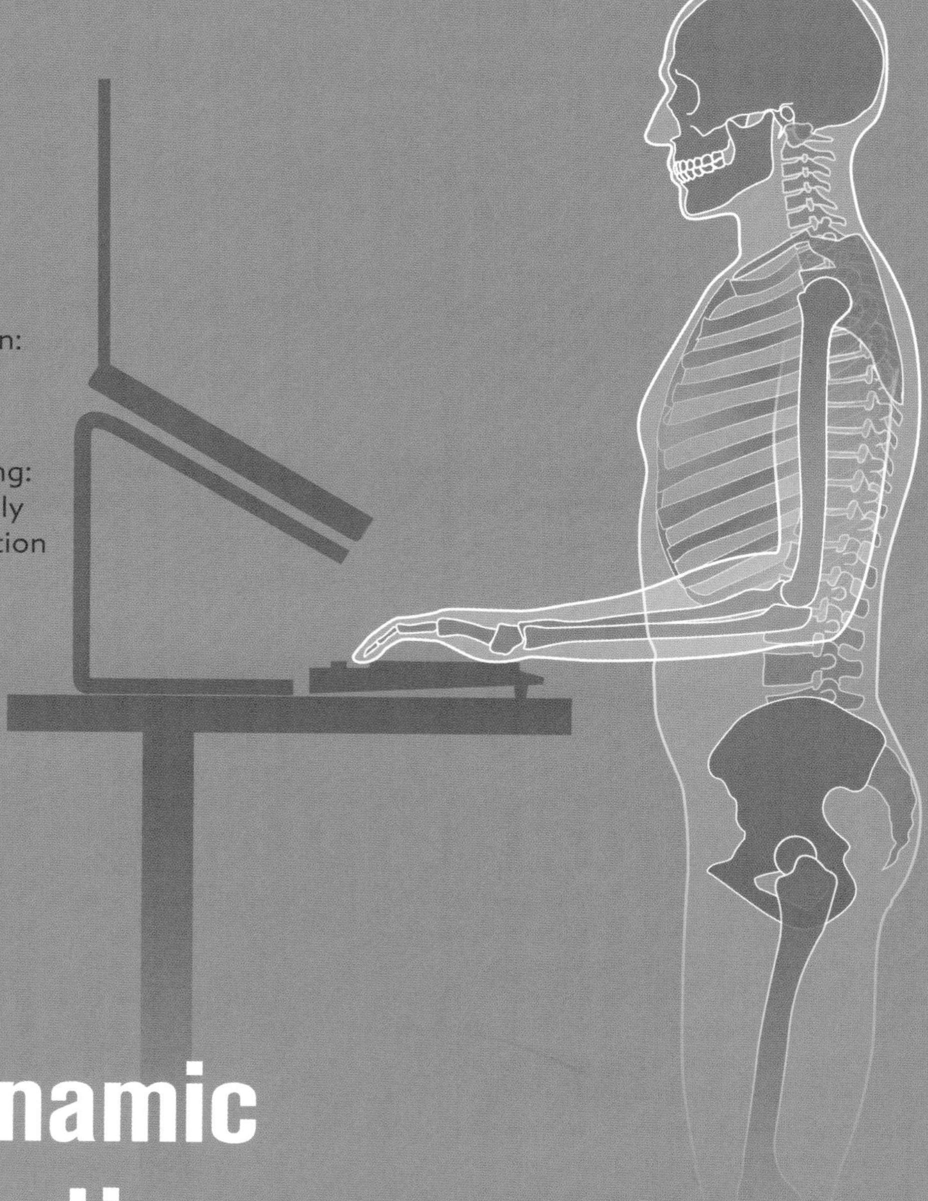

The Dynamic Workstation

Standing at work is a far better option than sitting because standing is a gateway to movement. But transitioning to a standing desk is a more complex process than just trading one piece of furniture for another. A successful standing workstation must be set up correctly, it must create opportunities for movement, and you must make the switch gradually to give your body time to adapt. We have broken this section into three chapters to help you make the transition a successful and lasting one.

First, we discuss the ergonomics of your standing workstation, from the proper height for your monitor and keyboard all the way down to the best type of shoes to wear. This advice is intended to create an environment that makes it easy for you to maintain solid body mechanics throughout your workday.

Second, because standing still like a statue all day is not the goal, we offer several standing positions that you can cycle through. Routinely changing your position keeps your muscles engaged, keeps blood flowing throughout your body, and makes standing more manageable throughout the day. Next, we cover a host of actual movements that you can perform to prevent some of the harmful side effects of inactivity, which is the true benefit of transitioning to a standing desk.

Finally, we provide a simple blueprint for safely and effectively transitioning from a sitting workstation to a standing one. This last chapter addresses the concerns that go along with making the seemingly dramatic switch to a standing desk.

Standing Workstation Setup

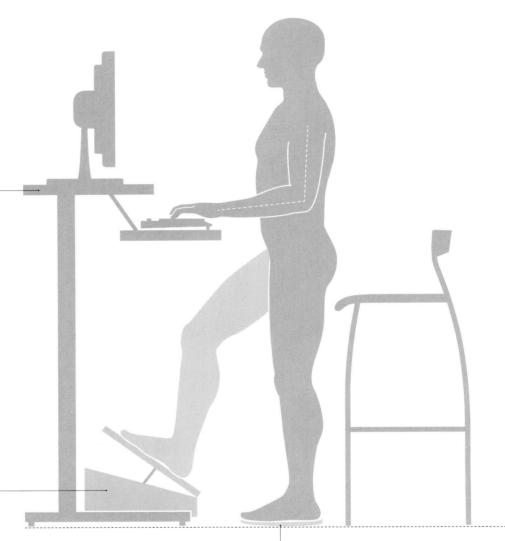

Desk:
The ideal standing desk allows you to position your monitor, keyboard, and mouse at the proper heights—heights that correspond to your now-good mechanics.

Footstool, Slant Board, Foot Rail, or Fidget Bar:
Resting your foot on an elevated surface automatically unloads many of the passive forces tugging on your spine, making standing more sustainable. Use a footstool, slant board (shown), foot rail, or fidget bar that reaches to about the middle of your shin.

Shoes:
Being barefoot is the best option, but it's unrealistic for most people. So the ideal shoe should make you feel as though you are barefoot. Consider working in a shoe that has a flat sole and just a little bit of cushion. Avoid high heels, restrictive dress shoes, and flip-flops whenever possible.

Monitor:

Position your monitor so that the top of the screen is aligned with your eyes. Consider tilting it upward so that you can see the entire screen better. You want your face to be about 18 to 30 inches from the monitor, which will allow you to see the whole screen without adjusting the position of your head.

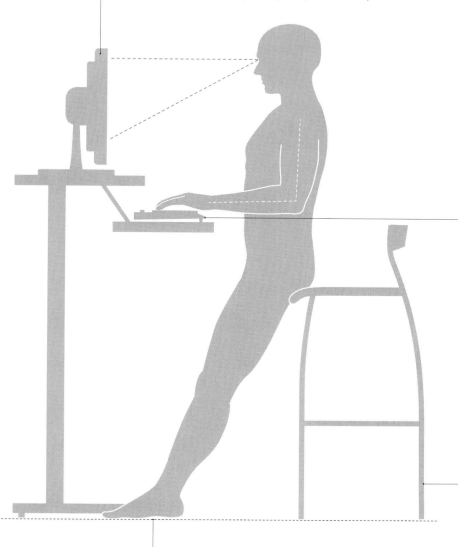

Keyboard and Mouse:

To find the right height and distance for your keyboard, go through the steps to reorganize your spine and bend your elbows so that your forearms are roughly parallel to the floor. Position your keyboard and mouse directly underneath your hands. Focus on keeping your shoulders organized and your elbows and wrists in clean, straight lines.

Stool:

The stool is used primarily for leaning against, not for sitting on. For this reason, you want a stool that has a hard seat with a squared edge and a wide enough base that it doesn't roll or fall backward when you lean against it. A simple bar-height metal or wood stool should work.

Floor:

If you work on concrete or another hard surface, consider using an anti-fatigue mat or wearing a shoe with a little bit of cushion. If you work on soft carpet, wear a flat shoe with little or no cushion or, better yet, work barefoot.

Standing Workstation Guidelines

Before we delve into the details of setting up a standing workstation, we want you to put down this book, stand up (if you aren't standing already), remove your shoes, and assume the organized spinal position. (If you need to, revisit the step-by-step guide on pages 82 and 83.) This is the primary position that you want to adopt while standing at your desk. Instead of changing your shape to fit your desk, you are going to change your desk to fit your optimal shape.

For years, we've seen people build elaborate standing desk setups that were based on really poor posture. For example, the monitor was too low, which forced them to look down. Or the desk was too high, which caused them to reach for the keyboard and put continual strain on their shoulders. While no desk setup can be reduced to an exact science, the key to happiness at work is to ensure that your desk supports and reinforces your improved mechanics. In a nutshell, a desk that fits your organized body makes your transition from sitting to standing and moving easy.

Misplaced Precision

We know and have great respect for many ergonomists and their work, much of which has significantly impacted and influenced this book. We have found, however, that some office ergonomic consultants suffer from what we call "misplaced precision." When Juliet was still practicing law, the firm's ergonomics consultant moved her mouse 3 millimeters to the right, gave her a curved keyboard, and said that she was now in a good ergonomic position. There was no discussion of posture or body mechanics, and she was still slouching like a shrimp in her chair. The problem is, there is no fancy keyboard, keyboard tray, or track ball mouse that can combat your poor sitting positions. Keep an eye out for misplaced precision.

Flat Shoes Are the Best Choice

This standard is not complicated but is potently effective. When you wear shoes, flat ones are your best bet. Flat shoes are an integral part of a functional standing workstation because they have a dramatic positive impact on posture and movement. And it is an irrefutable fact that the geometry the body was designed to support has the heel flat on the ground.

We've talked about how the majority of the kids in our daughter's elementary school class lost their perfect running mechanics during the course of their first-grade year, and part of the reason was the shoes they were wearing. The same holds true for adults. You want to cancel out any time spent in a shoe that shortens your tendons and connective tissues and kills off the range of motion that your feet are designed to wield. Our entire physical structure evolved with the foot bare, not confined to a shoe with an elevated heel. Do your fingers touch while your hand is at rest? Nope. Well, guess what? Your toes shouldn't be touching, either.

As described in Section 3, wearing shoes that don't support your biomechanics can be devastating. For example, if you wear heavily cushioned shoes at your desk, you are more likely to shift your weight onto the outsides of your feet as you slop around looking for a supportive position. Cushioned shoes also make you more prone to allowing your arches to collapse into the support of the shoes, impairing the mechanics of your feet and ankles. The arch of your foot is actually a non-weight-bearing surface. Think of any arch you see in nature or in architecture; nothing is holding up that arch in the middle. The same applies to your foot's arch; it doesn't need anything to hold it up. Your feet are the foundation upon which your spine is built. Setting your spine atop collapsed arches is like building a castle on sand.

High heels are an especially big problem. When a shoe fundamentally changes your natural, protective biomechanics, it puts you at increased risk for foot and ankle issues, low back pain, pelvic floor dysfunction, and more. And the first thing you do when you feel uncomfortable while standing in heeled shoes is to sit down! If that isn't enough to make you change your footwear habits, ask yourself if you would allow your child to stand on a 30-degree slope all day long. Assuming that the answer is a resounding "no," it's time to ditch your everyday high heels. If you are invited to the Oscars, we fully support you wearing some amazing stilettos. We simply recommend that you minimize your exposure whenever possible. The reality is, if you fail to address your footwear, you are setting yourself up for a future breakdown.

Flat shoes are much better for your biomechanics. A flat shoe has "zero drop"—meaning that the heel is not raised above the forefoot—and it should have enough cushioning that you won't injure your feet when walking across a gnarly gravel driveway. What you want to avoid, or wean yourself away

from, are shoes with the heels raised higher off the ground than the forefeet. Have a look at the shoes in your closet. We suspect that the vast majority have some sort of a heel. Most running shoes have heels over half an inch high, as do most men's dress shoes. That trend is changing, however, and now there are many companies offering flat *and* fashionable shoes.

The ideal is to flex your indigenous mechanics and go barefoot while standing at your desk, but we recognize that this is not realistic for anyone who doesn't work from home. If you can't work barefoot, choose the flattest shoe you can find with minimal padding. Remember the 1980s, when women walked to work in their comfortable shoes and switched to high heels once they got to the office? If you simply cannot give up your high heels, we recommend switching that model around: wear your heels to work and then stand at your desk throughout the day in flats or no shoes.

If you've been wearing heeled shoes your whole life, there is a good chance that you lack normal ankle range of motion, have compromised arches, and/ or have valgus knees (that is, you are knock-kneed). For you, the transition to flat shoes will take some time and investment. If you are exercising in heeled training shoes, we recommend a gradual transition to flatter and flatter shoes until you feel comfortable training in zero-drop shoes. For everyday wear, we recommend that you get a shoe that is as flat as possible, with minimal cushioning, and focus your attention on rebuilding your feet with Prescription 13 on page 334.

Here's a set of criteria to steer you down the right path. Your shoes should be:

- Flat-soled

- Adequately padded to protect the bottoms of your feet

- Nonrestricting, with enough space for your toes to spread out naturally

- Flexible enough that you are able to bend them

Take this list with you when you go shopping, try on the shoes that you're thinking of buying, and see if they pass muster.

The Floor Matters

Think about the story of Goldilocks and the three bears. Goldilocks tries all three of the bears' beds. One is too hard and one is too soft, but the third bed is just right, and that's where Goldilocks decides to rest. This same principle applies to the flooring at a standing workstation. The floor should not be too hard or too soft, but just right. The hardness of the floor will have a measurable impact not only on how you feel throughout your workday, but also on how well and how often you move. How hard is "just right"? Everyone has a different tolerance level, but we'd like to offer a few common-sense guidelines.

A friend of ours switched to a standing desk and then called us to ask our advice about why his feet were so sore at the end of the day. When we asked about his workspace setup, he reported that he wore dress shoes and stood on a concrete floor. We recommended that he switch to a shoe with a bit more cushion, and his foot pain went away immediately. You see, his setup didn't allow for enough movement of his feet during the day. Hard surfaces are okay; you just have to move your feet a lot more often.

If the floor beneath your desk is too hard, another option is to purchase an anti-fatigue mat. These mats are squishy enough to allow your feet to make small movements, but not so squishy that you can't find a stable position.

A floor that is too soft can make it challenging to find a stable foot shape. Try standing on your mattress for 10 minutes and prepare to be amazed at the ways your body deals with that squishy standing nightmare. The good news is that most carpets fall into the "just right" category.

The best advice we can give you is to monitor your setup when you first make the switch to a standing desk. If you're standing on a hard surface and your feet begin screaming at you an hour into your workday, try incrementally adding some cushion until the situation improves. Just make sure to check your position often and make adjustments as necessary.

Anti-fatigue mat

Flat shoes with a little bit of cushioning

A Footstool, Slant Board, Foot Rail, or Fidget Bar

Have you ever bellied up to a bar? Well, assuming that most of you have, you know that there is a rail at the base of the bar–that thing you put your foot up on while standing at the bar. It turns out that bar owners are pretty smart. They learned early on that if people had a place to rest their feet–on the rail–then they could stand and drink longer. "Bar height" makes more sense now, doesn't it? Bartenders were probably the first ergonomics experts on the planet. A standing-height drinking table that you can lean on, with a place to rest your foot? Genius.

Propping one foot up on a rail is the easiest way to remove the load from your lumbar spine (and reduce standing-related losses in alcohol sales). This pose—we call it the "Captain Morgan," for obvious reasons—automatically puts your pelvis in a better position, which means that you don't have to work as hard to stay organized. Simply put, a foot rail makes it easier to stand comfortably for longer periods. In fact, having some kind of foot support is so critical that we don't consider a standing workstation complete or acceptable if it doesn't give you a place to rest your foot.

So how do you incorporate the requisite elevated foot into your standing workstation without relocating the office to the local watering hole? The cheapest and easiest method is to use your stool, a box, or an abandoned chair. The height can vary, but when your foot is propped up on it, your knee

should be no higher than your hip. Some people like a little less or a little more elevation. Experiment for yourself, and don't be afraid to mix it up.

A slant board is another excellent option. It is one of our top recommendations because it gives you more choices. It's similar to a simple platform, but instead of a flat top, it has an incline. This incline allows you to stretch your calf and ankle while in the Captain Morgan Pose (Option 1). You can also stand with both feet on the slant board (Option 2), which improves your balance and allows you to work on ankle and calf mobility without disrupting your work flow. Plus, you can use the slant board as a footrest when you're leaning against or sitting on your stool (Option 3).

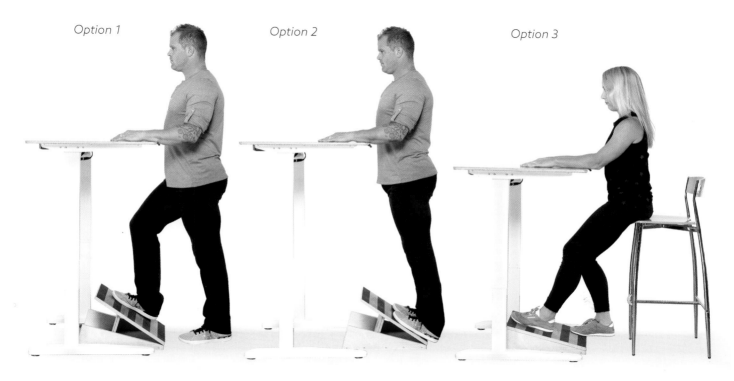

Option 1   *Option 2*   *Option 3*

Finally, you can incorporate a relatively new design innovation and get yourself a swinging foot bar, or fidget bar. This is essentially a bar on which you can rest your foot that also swings back and forth. When we supplied our daughter's all-standing/moving school with standing desks, we made sure that the desks had these swinging bars. The fidget bars allow the kids to engage in constant small motion throughout the day that doesn't disrupt class. The students and teachers report that the bar is their favorite part of the standing desk setup.

If 450 school-age children can make the move to a dynamic standing work environment, so can you. Keep an open mind and be creative. There are plenty of options that will help you achieve your goals of increasing your movement throughout the day and supporting your working postures.

The Desk Itself

A decade ago, it was nearly impossible to find a standing desk at all, let alone an affordable one. Today, dozens of companies are making adjustable and fixed-height standing desks. Standing options are becoming more and more affordable as they become more commonplace.

There are also many innovative for-purchase and do-it-yourself solutions for converting a sitting workstation to a standing one. One company even sells a $25 cardboard standing desk converter. Still too fancy? Grab an Amazon box and get started. Below are some of our favorite standing desk hacks that people have sent to us.

If you are handy and enjoy DIY projects, you can easily create your own standing workstation using materials that you probably have lying around the house. We know people who have used cinder blocks and common building materials to raise the height of their desks to match their own height and proportions. We've even seen people extend the height of their desks by sliding PVC pipe over the legs. Pam, one of our best friends, works at her

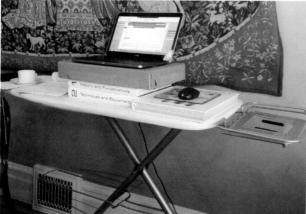

ironing board standing desk whenever she works at home. The point here is that you don't have to spend thousands or even hundreds of dollars to convert to a standing workstation. If you're hunting for ideas, simply type "standing workstation conversion" into your search engine and a host of solutions will pop up.

Adjustable vs. Fixed-Height Standing Desks

As you've probably gathered by now, the key benefit of a standing desk is that it creates a movement-rich environment in which you can easily fidget, switch positions, and move around. There are two types of standing desks: adjustable "sit-to-stand" desks and fixed-height desks. When we say "fixed-height," we mean that the desk will adjust to your height, but it won't allow you to sit and work. Each type has its pros and cons, depending on your specific needs and personality, but we much prefer fixed-height desks.

An adjustable desk is the more flexible of the two options. With the push of a button or the turn of a simple crank, the desk transforms from standing to sitting height. This feature is appealing because many people seem to like doing email or talking on the phone while standing but struggle to do more focused tasks, like writing, unless they're sitting. There are several important drawbacks to adjustable standing desks. The first is price. Adjustable desks typically cost in the $1,000 to $4,000 range—enough to scare any employer who receives a standing desk request from an employee. Second, research out of Cornell University on sit/stand workstations shows that people tend to stand for the first few months (perhaps because of the novelty) but eventually lower the desk, sit back down, and never get up again.[1] An adjustable desk is nice in that it's easy to make small changes to your standing work height given your current task. But treat that down button with the respect that you'd afford a loaded gun. If you purchase an adjustable desk and revert to sitting, you've just traded one sitting desk for another, probably with a big price tag.

Fixed-height standing desks are much more affordable but lack the flexibility of adjustable desks. This lack of flexibility is actually a feature we like, so we recommend fixed-height desks to most of our clients. Having a stool gives you a place to rest when you're tired, but you have no option to revert to the dreaded sitting-all-day position. We vote for taking choice out of the matter. Around our house we have a saying: if we don't want to binge on chocolate chip cookies late at night, we can't have piles of chocolate chip cookies lying around. If you think you'll revert to sitting if you have the option, we couldn't agree more—you will.

Desk Height

The most important consideration with a standing desk is height. The majority of fixed-height standing desks are 40 or 42 inches tall. According to Dr. Mark Benden, professor and director of the Ergonomics Center at Texas A&M University and author of *Could You Stand to Lose? Weight Loss Secrets for Office Workers*, this height allows 90 percent of the healthy adult population to work comfortably while standing.[2] Before you run out and purchase one of these desks, though, make sure that you actually fall within this 90 percent. If you don't, an adjustable desk might be a better option for you.

First, consider how you use your desk. The goal is to find the ideal desk height for the type of work that you perform. If you spend most of your time using pen and paper, you'll want to measure from the floor to the surface of the desk. If you primarily use a keyboard and mouse, measure from the floor to the keyboard tray. If your work involves a combination of both, you'll probably want to forgo the keyboard tray and place your keyboard right on your desktop so you can easily switch back and forth between typing and writing or drawing.

Next, go through the bracing sequence on pages 82 and 83. With your posture corrected, bend your arms so that your forearms are parallel to the floor. Your desk or keyboard tray should be at elbow height. But don't be dogmatic about it. If you work at the suggested height for an hour and think you will feel more comfortable if you move it up or down a little, trust that instinct.

*How to Set the
Height of Your
Standing Desk*

The Leaning Option: A Stool

A stool is a critical part of your standing workstation, but not all stools are created equal. The goal of having a stool is not to create a tall sitting desk but to give you a surface that you can occasionally lean on or use to prop up one leg.

In our experience, a stool with a hard or semi-hard surface, a flat pad, and squared edges is best because it grabs the edge of your glutes and prevents you from slipping off. It also creates the best surface for leaning. A stool with curved or soft edges will be uncomfortable to lean against and will probably squirrel away from your desk (as will a stool that has wheels).

A stool with a flat pad is also easier to use for performing basic body maintenance on the job. For example, you can prop your foot up to mobilize your hamstring or lay your leg across the stool to open up your hip, as shown opposite. In addition to giving you options for moving, a stool with a hard surface will help you maintain a better temporary seated position. As you will learn in Section 5, having a hard surface to sit on is fundamental to sitting in a good shape.

Most people don't know that the material used in Herman Miller chairs was originally developed to prevent pressure ulcers caused by sustained sitting. Anyone who is chiefly confined to a sitting environment like a wheel-chair will tell you that they obsess about their seating surface. The problem with relying on advanced material technologies if you don't have to (like in your optional desk chair) is that it facilitates sustained pressure on tissues that were never intended to bear weight. A hard surface, on the other hand, allows for great mechanical feedback; you get uncomfortable, which reminds you to move on to your next shape before too long.

It's worth mentioning here that there are several stools, stands, and dynamic seating options that you can perch on or lean against while remaining in an upright position. Most of these seats are specifically designed for adjustable desks. Although we offer a specific recommendation, we want you to know that you have options. As long as the seat promotes standing and accommodates upright leaning and perching, it will get the job done.

What about a stool with a backrest? Well, the backrest encourages sitting and slumping. Why use the beautiful machinery of your body to support yourself when your stool will do it for you? When you do decide to sit on your stool, the lack of a backrest will force you to maintain an upright position and keep your trunk engaged. If your stool has a backrest like the one shown opposite, resist the temptation to use it.

It was only a short time ago that lifting belts were common in industrial work environments. The thinking was that if workers kept their vulnerable midsections braced with a belt all the time, the number of lumbar lifting

injuries would decline. This makes perfect sense, right—extra support for people who lift heavy things for a living? As you might imagine, it backfired for precisely the same reason that we don't advocate using a backrest for support. As soon as workers began to use the belts as the primary mechanism for stabilizing their spines, their trunk musculature basically went on vacation, creating a downward spiral of core de-conditioning, and injury rates went through the roof. The same thing happens to your trunk musculature when you decide to let a chair act as a surrogate for the stabilizing system of your spine.

As with your desk, the height of your stool is very important. The ideal height is roughly the same as your inseam. Another, perhaps easier way to get this measurement is to pick a stool with a seat that just reaches your butt. When you're leaning against the stool, you want the crease of your butt cheeks positioned against the edge of the seat, which provides equal amounts of comfort and grip. If the seat is too low, you won't be able to reach your keyboard (assuming that you have a fixed-height desk). If the seat is too high, you won't be able to lean against it effectively. When the height of the seat is just right, your lean will look a little like sitting and a little like standing, yet will remove some of the work from your legs. If you've ever found yourself in a crowded European train or on any form of public transportation, you'll recognize that you can basically lean forever.

Your Monitor

Simply plopping your monitor on top of your desk is not going to create the optimal setup for your body. If you are at the point where you are placing a monitor on top of your new standing desk, you are already crushing life, but we have seen too many beautiful necks lost to wretched monitor placement.

Determining the ideal height for your monitor is pretty easy. Organize your body into your home base spinal position, focus your gaze straight ahead, and then position your monitor so that the top of it is at eye level. If your monitor is adjustable, it may be beneficial to tilt it upward slightly so that you can see the entire screen. This keeps your head aligned with your neck and trunk. Just because you *can* orient your head downward for hours on end doesn't mean that you should. The first step in stopping headaches and neck pain related to poor head positioning is to control and change the environmental demands.

The difficult part is not determining the proper height for your monitor but implementing it. If your standing desk is at the right height and has a keyboard tray positioned below the surface of the desk, everything may work out perfectly. Your forearms will be parallel to the floor while you type, and your eyes will be focused straight ahead. However, if your keyboard and mouse are sitting on the surface of the desk, chances are your monitor will be too low. Several companies sell standing desks that offer height-adjustable monitor arms, but be forewarned: this option often requires some assembly.

If the standing desk you're considering doesn't have this option or you are constructing a makeshift standing desk, you'll need to find a way to elevate your monitor to eye level. The simplest and cheapest method is to position it on top of a stack of books or a small box, but a monitor stand on your desktop looks far more attractive and is probably more appropriate in an office setting. Still, we love a makeshift monitor setup. It screams, "At least someone around here cares about their neck!"

If you have a fixed-height standing desk, we recommend setting your monitor height to your standing position rather than your stool-based leaning position, which tends to be a few inches lower. Acrobatic divers have always been champions of good ergonomics: they know that where the head goes, the body follows. If your screen is too low (as it would be sometimes if you set your monitor height to your leaning position), you will look down. The next part of the body to follow will be your neck, followed by your trunk. How about this for a monitor rule? See your future, be your future. Or, in the case of a too-low computer screen: don't look where you don't want to go.

We aren't sticklers for how far away your screen should be, but our friends who obsess about these things offer some basic guidelines. When it comes to the distance between your eyes and the monitor, the key is to be able to see the entire screen without having to move your head. The exact measurement depends on the size of your monitor, your eyesight, and what you're working on. For some jobs you may need to be closer to the screen, and for others you may need to be a bit farther away. If the monitor is too close, you might strain your eyes, and if it's too far away, you might have to crane your neck forward to see the screen better. The general recommendation is to position your monitor between 18 and 30 inches from your eyes, but these numbers are somewhat arbitrary. Do what's most comfortable for your body type, your job, and your setup.

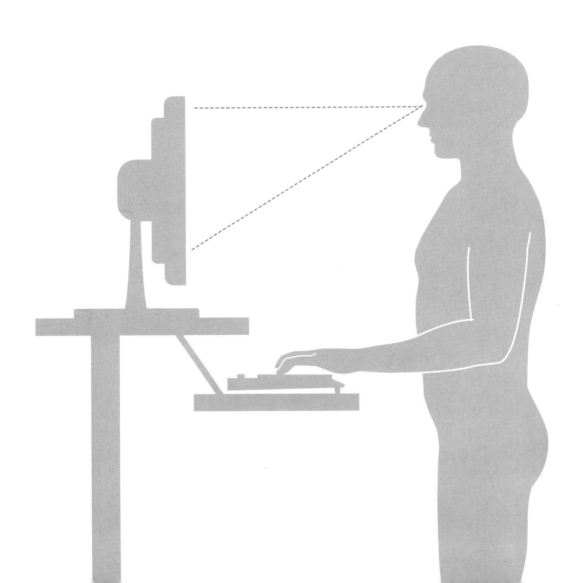

Your Keyboard and Mouse

The primary consideration with your keyboard and mouse is to have your forearms about parallel to the floor while you work. On page 149 we gave you a method for determining the right height for your desk, which you should reference if you want your keyboard and mouse to sit on top of your desk. If you plan to use a keyboard tray, then the height of your desk matters only for your monitor—it's the height of the tray that matters for your arm positioning. Use the same method to find that height: Establish an organized standing position and bend your elbows about 90 degrees so that your forearms are parallel to the floor. Now set your keyboard tray to elbow height or slightly lower, depending on your personal preference. Some people find it more comfortable to have their arms at 90 degrees, while others prefer a slightly more open angle. Your elbow height will ultimately dictate your ability to keep your hands in line with your wrists and elbows.

We also recommend standing or sitting close to your keyboard, which enables you to achieve the ideal upper body position—externally rotated shoulders, elbows in tight to your body, and wrists aligned with your elbows and shoulders. Because this anatomical position mimics the lotus position in yoga, we like to call our interpretation the "Eastern Keyboard Approach." If you're farther away from your keyboard and mouse, you'll have to reach your arms forward, which causes your elbows to flare outward, your shoulders to internally rotate, and your wrists to collapse. Working in this position for hours on end, day after day, can lead to thumb and wrist problems and is a mechanism for repetitive stress injuries like carpal tunnel syndrome.

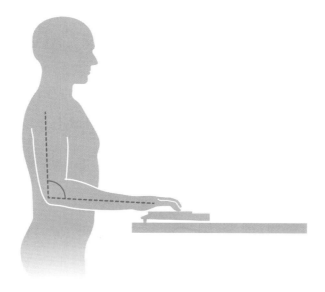

Over-Glow: How Your Computer Is Affecting Your Eyesight and Sleep

According to the American Optometric Association, if you spend two or more consecutive hours staring at a computer screen every day, you are at greater risk of developing computer vision syndrome, or digital eyestrain.[3] That's right, there's an eye syndrome associated with too much computer time. Similar to carpal tunnel syndrome, computer vision syndrome is basically a repetitive stress injury for eyes. You can develop a host of symptoms ranging from eye fatigue, eyestrain, and burning and itchy eyes to sensitivity to bright light, blurred vision, headaches, and even neck and back pain. Playing video games, watching TV, and using a tablet can cause similar eye problems. As with other repetitive stress injuries, the onset of symptoms occurs when you carry out the same motion over and over again.

But why is looking at a glowing screen so bad? There are a few reasons. For starters, your eyes have to constantly adjust between staring at the bright computer screen to reading a paper to gazing out a window. Add to that the glare, flicker, and changing images on the screen. All this requires a lot of effort from your eye muscles, and although there is no evidence that it causes long-term damage, short-term ramifications are common. The good news is that there are solutions.

- **Take a break:** Just as you need to take a break to move, you need to take a break from looking at your screen. Our physiology just wasn't set up to stare at a fixed distance for 40-plus hours a week until the age of 65. Really concerning research around sustained fixed-distance gazing comes from people investigating the physiological effects of long-term, close-quarters confinement in prison populations. Because prisoners spend most of their time in small jail cells, their eye function degenerates. Are your eyes in self-imposed solitary confinement? This is where the 20-20-20 rule comes into play. Take a break every 20 minutes and look at something 20 feet away for at least 20 seconds. Many people don't blink as often when working on a computer, so when you do take a break, try to refresh and moisten your eyes by blinking. If you follow the guidelines outlined at the beginning of this book, you're already taking a short movement break every 20 to 30 minutes, so use this time to look at something other than your computer screen, too. Take a quick walk outside or stare out a window.

- **Reduce glare and brightness:** Adjusting your monitor settings and reducing glare may help. Make sure that the light is evenly distributed around your desk, and customize the brightness, contrast, and text size on your computer to find the best settings for your eyesight.

- **Follow the workstation setup guidelines:** Make sure your monitor is positioned so that you don't have to lurch your head forward or strain your eyes to see what's on the screen. Again, 18 to 30 inches is the general recommendation.

You've probably heard that being exposed to light at night, especially the blue light emitted by electronics, is bad for your health. Why? It's all about your circadian rhythm, which is your body's biological clock. Being exposed to light (specifically blue light) at night disrupts your body's natural sleep cycle. As a result, your sleep is compromised. And when your sleep is compromised, everything is compromised.

Blue light produced by electronics is especially bad because it causes the body to suppress the release of melatonin, a hormone that helps regulate sleep and wake cycles. When it gets dark, your body releases melatonin, which tells your brain that it's time to relax and go to sleep. So, when you're up late working on your computer, your body never gets the signal that it's time to sleep. If you have a hard time getting to sleep at night, this may be the problem. Fortunately, there are some simple remedies.

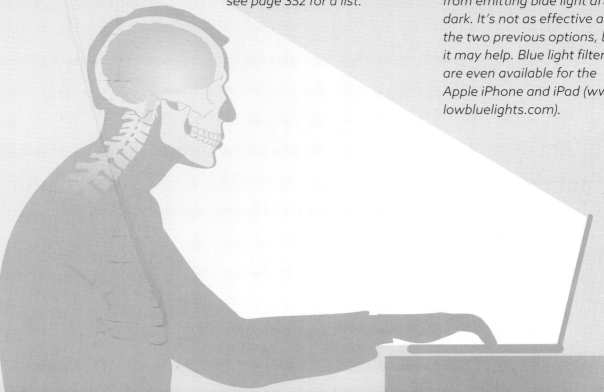

Blue Light and Sleep

- **Avoid using electronics after dark:** *Turning off all unneeded lights in your home and avoiding bright screens for two to three hours before bed is the most straightforward solution. However, that is not realistic for most people.*

- **Wear blue light–blocking glasses:** *Studies have shown that wearing blue light–blocking glasses tricks your body into producing the same amount of melatonin that it would if it were dark, even when using electronic devices.[4] And no, you don't have to look totally dorky doing it. There are some quite fashionable options available—see page 352 for a list.*

- **Install software that reduces blue light emissions:** *Another way to block blue light is to install a software program called f.lux (www.justgetflux.com) on your computer. f.lux adjusts your computer display according to the time of day. The screen is bright during daylight hours, but f.lux prevents your monitor from emitting blue light after dark. It's not as effective as the two previous options, but it may help. Blue light filters are even available for the Apple iPhone and iPad (www.lowbluelights.com).*

If you find that you are having trouble maintaining the Eastern Keyboard Approach, you may want to consider an ergonomic (curved) keyboard, which better supports your indigenous mechanics. The traditional flat keyboard is a wonderful example of how comfortable we've become with conforming our bodies to our technology. In an ideal world, you'd be able to split your keyboard in half so that it would work at the distance between your hands when your elbows are bent at your sides and your forearms are straight out.

When it comes to typing mechanics, keeping your wrists aligned with your forearms is critical, as is taking frequent breaks to move your wrists and hands. The crucial takeaway is that your wrists should not be resting at the base of your keyboard, at the bottom of your laptop, or on your mouse pad. Instead, your hands should be supported at your shoulders with your wrists free-floating or lightly brushing the keyboard. Resting your wrist can compress the nerves and blood supply to your hand, as it compresses the tissues that run through your carpal tunnel, the normal bony tunnel in every wrist. In addition, your anchored hand cues your body to stop creating a stable position upstream at the shoulder.

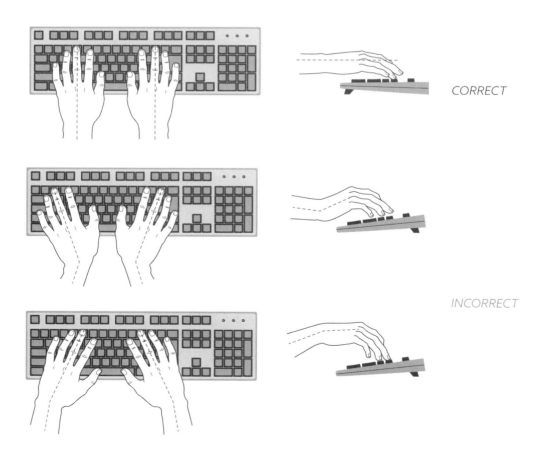

CORRECT

INCORRECT

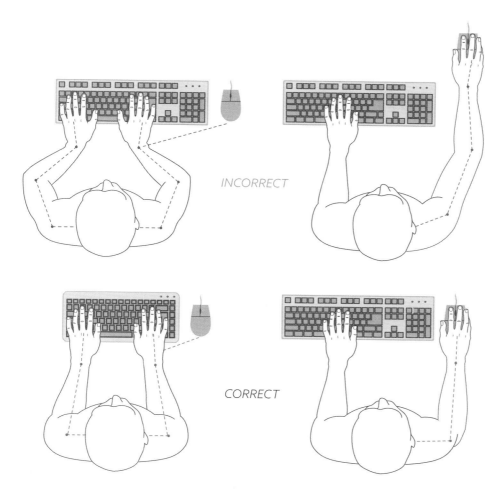

INCORRECT

CORRECT

These same principles apply to how you move and scroll with your mouse. Again, the goal is to keep your wrist aligned with your forearm, positioning the mouse close to your keyboard and body. Avoid resting your wrist on the mouse pad or desk, and avoid moving the mouse with your wrist alone. Instead, move from your shoulder. For example, if you have to move your cursor across the screen, do it with your shoulder, which you'll notice moves your entire arm.

If you position your elbows at your sides, your forearms parallel to the floor, and your hands like they were performing some karate chopping, this is where your mouse should be. Using your mouse with your hand inside your elbow is a recipe for shoulder and neck dysfunction. Keeping the mouse even with your forward-pointing forearm or even slightly outside it utilizes your body's natural mechanics and connective tissues to support your whole arm. With the mouse oriented in this position, you are better able to maintain an organized, stable shoulder and neck.

Mouse mechanics are a low-effort way to make your world fit your anatomy rather than force your anatomy to fit your world. You know that knot on your mouse-using side that goes from your shoulder blade to the back of your neck? This ropy piece of beef jerky is the calling card of "mouse in the wrong place" syndrome. If you want it to go away, you are going to have to give it a break by improving your mouse hygiene.

Lastly, because we know that typing in the perfect, anatomically sensitive position may not be possible all the time, we also recommend doing some basic maintenance by rolling out your forearms, wrists, and hands every hour, or more often during heavy typing and mouse usage—see Prescription 7 on page 290.

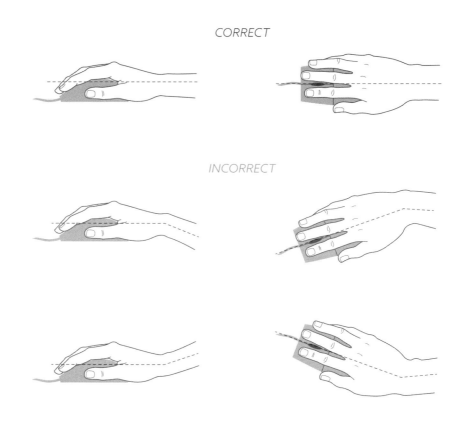

CORRECT

INCORRECT

The Laptop Workstation

Using a laptop makes it a bit trickier to set up an optimal standing workstation because the keyboard and monitor are a single unit. If your laptop keyboard is at the height that we recommend for a regular computer, your monitor will be far too low. Remember, your head follows your gaze, so you will ultimately assume the dreaded rounded-back position. If your monitor is at the correct height, it will be impossible to keep your forearms parallel to the floor.

There are two solutions to this problem. The better option is to purchase either an external monitor or an external keyboard, which allows you to position both your keyboard and your monitor at the proper height.

INCORRECT

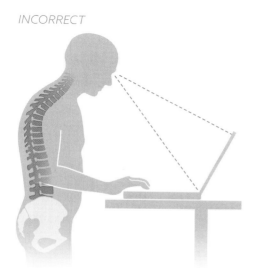

CORRECT

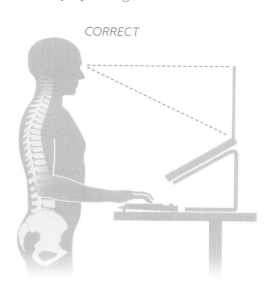

The other option is to split the difference by positioning the top of the laptop screen at roughly neck height. It's not ideal for extended periods but shouldn't cause you too much concern when used intermittently.

The worst laptop ugliness is to sit in a soft chair, hunched over your tiny body-crushing device. Yes, we are referring to every business traveler we've ever seen. Ever.

The Active Workstation: Creating a Movement-Rich Environment

A standing workstation enables you to burn more calories, activates more muscle groups than sitting, and increases blood circulation. Standing also reduces your risk of cardiovascular disease, obesity, diabetes, and some cancers. But when you are accustomed to sitting, standing all day can be strenuous. This is especially true if you are overweight or have been deskbound for the majority of your adult life. The fact that standing is strenuous is a sign that your body is out of whack, because the human body is designed to be upright and moving. That said, we appreciate that the transition from sitting to standing can be challenging at first. The simple key to making the transition smooth is movement. Before you have a panic attack and scream, "There is no way my boss will let me run laps around the office all day," understand that movement comes in all shapes and sizes—literally.

Three general types of movement will give you the most benefits from your standing workstation:

1. **Shifting positions:** This, the least dramatic form of movement, doesn't require you to leave your computer or interrupt your work. It's as simple as switching from one braced neutral standing position to another every two to three minutes, or whenever your body sends you a signal to move. In this chapter we cover three base positions that you can cycle through.

2. **Movement breaks:** Even if you change your position often, you still need to infuse your workday with actual movement that forces your body into different shapes. The goal is to bury two-minute movement breaks into your workday. This practice will get blood circulating to areas that have been idle and poorly perfused, most notably your neck, arms, wrists, and fingers. Take these short breaks every 20 to 30 minutes.

3. **Mobility breaks:** Two-minute body maintenance (or mobility) breaks should be interchanged with your movement breaks. Instead of moving, try performing one of the techniques from the sample routine outlined at the end of this chapter, or pick one from Section 7. Some of these techniques will take you away from your keyboard momentarily, but there are an equal number that you can perform while at your workstation.

Shifting Positions

Regardless of how well you organize your body at your standing desk, remaining in the same position for hours on end just isn't ideal. To avoid the stillness trap that you experienced at your sitting desk, be conscious of shifting your position as often as possible to create movement. A nice way of saying this is that your best position is probably your next one.

There are three basic positions that you should cycle through: Stable Standing, the Captain Morgan Pose, and the Upright Lean.

Stable Standing:
While Stable Standing is an ideal position for your body, it is challenging to maintain for long periods. Standing well in one place for a long time is ultimately uncomfortable and potentially strenuous. And standing in one position without moving doesn't respect the movement-based design of the human body. For this reason, you probably will spend the least amount of time in this position. Most people use Stable Standing as a transitional position between the Captain Morgan Pose and the Upright Lean.

Captain Morgan Pose:
Elevating one foot off the ground automatically orients your pelvis into a more neutral position and reduces potential tension and strain in your lower back. This makes the Captain Morgan Pose one of the best positions to assume while working at a standing desk.

Upright Lean:
This isn't technically standing, but it's not really sitting, either. It falls somewhere in between. The Upright Lean shares many of the same traits as standing in that your hips are open, your posture is upright, and your legs are straight. But it's also like sitting because you're resting your butt on the edge of a stool, effectively taking weight off your legs. If you have to work in front of a computer for eight hours a day, this is where you will likely spend the majority of your time.

Like the baseline movements outlined in Section 3, consider Stable Standing, the Captain Morgan Pose, and the Upright Lean to be your foundational positions. Once you can effortlessly shift between these positions, start adding new ones to your repertoire—like the ones we offer on pages 166 and 167. In the beginning, the key is to maintain a neutral body and shift your position as often as possible. That process is a lot easier when you have only three positions to manage. Feel free to create your own work posture language over time; just remember to respect your spine.

As you already know, your body does a pretty good job of letting you know when it's uncomfortable. Sometimes that discomfort registers as pain, and other times it manifests as body agitation. This is not unique to standing. If you were to film someone at a sitting desk for an entire eight-hour period, chances are you would see him assume all sorts of positions, such as slouched forward, overextended, legs crossed, and legs propped up on the desk. He assumes all these seemingly random positions because his body is screaming at him to move. But your movement options are limited when you are seated. As you are now aware, standing opens up a lot of new possibilities. Although the three baseline positions aren't dramatically different, they provide your body with just enough movement to keep your joints, connective tissues, and muscles from barking like a roomful of pent-up Chihuahuas.

The trick to dynamic standing is training yourself to default to one of these positions when your body sends you a signal to move. When you're first starting out, you will catch yourself in suboptimal positions throughout the day, like hitching over onto one hip. The only way to rid yourself of this habit is to correct your mechanics whenever you notice them slipping. Because your posture tends to slip the most when you are engrossed in a task, you might want to consider giving your coworkers license to call out your bad posture when they see it. We have this policy in place at our office, and we routinely correct each other's posture at work. And it certainly holds true for our daughters, whose job is to catch us in compromised shapes. Moving well is a skill, and skills take time to develop. Did you know that a child has to perform a task roughly 10,000 times before she acquires that new skill? That's a lot of practice.

Your body's signals will dictate how often you switch between these three stances. If your feet ache, your lower back gets tight, or you feel discomfort, it's time to switch. We've observed people who were able to stand in a static position for an hour or more before discomfort kicked in, but in those cases we concluded that they were simply too engrossed in their work to notice the signs. If you are such a person, we recommend making a mental note to switch every two or three minutes. You can even set an alarm on your phone as a reminder at first, until moving frequently becomes second nature. If you can keep moving at your desk throughout the day, you are accumulating a ton of NEAT movement without leaving your workstation. As we discussed in the introduction, that type of movement can literally mean the difference between being obese and maintaining a healthy weight.

There is no prescribed order or progression to these three foundational positions. Transitioning from Stable Standing to the Captain Morgan Pose is just as effective as transitioning from Stable Standing to the Upright Lean. The only rule is to do what feels comfortable.

When you're new to a standing workstation, it's quite possible that at some point in your day, none of the three positions will feel comfortable. That's when you should give your legs a break and sit on your stool. Or, even better, step away from your computer for a moment and perform a two-minute movement or mobility break.

Again, it's important to emphasize that you are not limited to these three positions. We encourage creativity, but we highly recommend starting with them if you are new to a standing workstation. All three positions make it easy to maintain an organized body, and in the beginning it is far more doable to commit to three positions than twenty. As long as the positions you assume adhere to your body's organizing principles, they should serve you well.

Guidelines for Shifting Positions

Shift as often as possible: *We often say that your next position is your best position, which is really just a cue to move often. To avoid sitting still, shift your position whenever you begin to feel uncomfortable or get the urge to move. This might be every 10 seconds or every minute, but you should avoid being in any one position for long stretches.*

Set a timer: *Most people shift positions intuitively, every few seconds or every few minutes at the most. But some people get so engrossed in their work that they forget to move. If you're one of those people, set a timer to go off every three minutes. This might prove distracting at first, but it won't take long for the habit to sink in.*

Change the position of your stool: *To encourage shifting, change the position of your stool often. Stand with your stool behind you, position it to the side of your body, or place it in front of you. Each stool position will force you to adopt a new shape. Here's one way to implement this strategy: every time you return from an every-30-minutes movement break, change the orientation of your stool. For example, for the first half hour, stand with your stool behind you so that you can lean, and then, after your break, move it in front of you so that you can rest your foot on it.*

To give you an idea of how to shift positions as well as give you some alternatives to choose from, here are some functional standing positions:

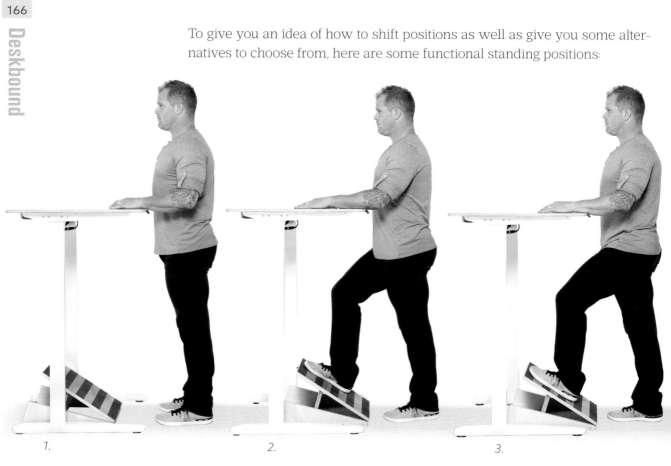

1.   2.   3.

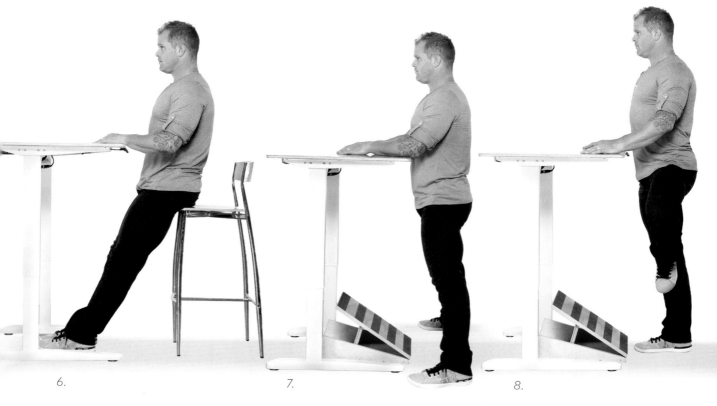

6.   7.   8.

4.

5.

9.

10.

Movement and Mobility Breaks

Constantly shifting your position while at your standing workstation will do wonders to prevent achy joints and muscles, but it simply isn't enough to maintain a healthy body. To avoid all the negative ramifications of being idle for long periods, you must add actual movement to your workday. Don't worry; we're not talking about dramatic exercises like burpees or weighted squats. Movement comes in many forms: taking a quick stroll around the office, rolling your wrists or shoulders, or squatting down into a chair several times. In Section 7, we offer a plethora of beginner to advanced mobility techniques that you can safely perform at your workstation without turning too many heads. Be creative! These are just suggestions to inspire you to move.

Movement breaks should be at least two minutes long and should be performed every 20 to 30 minutes whenever possible. This might seem like a lot of breaks, and you might be worried about how your coworkers or boss will perceive your time away from your desk. But moving is not just good for your body; it's good for your mind, too. Many studies have shown that the average worker "works" only three hours a day. The other five hours are spent procrastinating, chatting with coworkers, or staring mindlessly at a wall. We believe that the sedentary nature of many jobs is to blame. The research on brain function is clear: when the body is inactive, and especially when it is sitting, the brain cannot be fully engaged and it is much more difficult to stay focused. (See the sidebar "The Movement Brain" on pages 22 and 23 for more statistics.)

There is also an increasing amount of evidence showing that moving more throughout the workday enhances productivity and leads to better time management. For example, the footwear company New Balance piloted a program in which it encouraged roughly 750 employees to integrate some form of physical activity into their routines every 30 minutes. Of the 239 people who completed a post-program survey, 53 percent said that their level of physical activity at work had increased, and 42 percent reported heightened engagement and concentration.[5] Moving more can even help you manage work-related psychological stress.[6] Notably, tech giant Facebook reported that since incorporating standing desks into its offices, employees feel more energetic throughout the day.[7] And research has shown that standing may encourage more creative and collaborative group work.[8]

The key is remembering to take these breaks. If you have been desk-bound for a long time, you may revert to your old pattern of not moving enough. Luckily, there are a host of helpful tools that you can utilize. The simplest is a basic timer. We recommend setting it to go off every 20 to 30 minutes. When it beeps, finish up what you're doing and take your movement/mobility break. There are also more comprehensive tools, such as programs that will time-out your computer screen every 30 minutes. Here are some options worth considering:

- Focus Time activity tracker and timer (iOS) (http://focustimeapp.com/)

- Marinara productivity timers (Web) (www.marinaratimer.com)

- Stand Up! work break timer (iOS) (www.raisedsquare.com/standup/)

- Time Out break reminder tool (Mac) (www.dejal.com/timeout/)

- Tomighty desktop timer (Mac/Windows) (www.tomighty.org)

To give you an idea of how to implement movement and mobility techniques into your workday, we've drawn up a sample routine that you can perform in your workspace. If you're deskbound for eight hours and you're taking a two-minute break every half hour, you would perform one complete set of the movement and mobility techniques outlined on the following pages. You don't have to complete the sample routine in this exact order; you can and should alternate between movement and mobility sequences based on your personal preferences. Moreover, you don't have to perform all the movement and mobility techniques offered. Even doing five of them is better than nothing.

We encourage you to develop your own routines using techniques and movements that are specific to your body's needs and are appropriate for your work environment. In other words, you're not limited to what's included here. In fact, you shouldn't limit yourself to what's featured in this book. We know people who perform push-ups and bodyweight squats and go for short walks during their breaks. No movement is off-limits. The key is simply to move and mobilize.

Sample Movement and Mobility Routine

Break 1: Neck Movement

The neck is one of the first areas to get stiff when you're working at a desk. To combat stiffness, tension, and pain, simply move your head in all directions—tilting your head up and down and side to side and dropping your ear toward your shoulder—periodically throughout the day or anytime you feel tension building. You can spend time in each position or seamlessly transition from one position to the next.

Break 2: Wrist Roll

People go hours without lifting their hands from the keyboard, especially when they're engrossed in work. It's no wonder that so many of us suffer from hand-related problems. If your hands have been stuck in one position, get some movement into the area by doing circles with your wrists. Complete 10 full rotations with your hands moving in a clockwise direction, then do 10 in a counterclockwise direction. Switch back and forth like this for a minute or two. You can also shake out your wrists as if you were trying to air-dry wet hands and/or open and close your hands.

Break 3: Hip Opener (Split Squat)

The split squat is great for opening up the hips and activating muscle groups in your lower body. To begin, get into a lunge stance with a slight bend in your knees, keeping the glute of your rear leg engaged. From here, lower your rear knee to the ground, dropping it straight down. Focus on keeping your rear glute engaged, your lead shin as vertical as possible, and your torso upright and neutral. Ten split squats on each leg (right leg forward and then left leg forward) is a good number to aim for.

Note: You don't have to perform the split squat movement. Simply getting into the lunge position gives you a nice stretch through your hips. Try to accumulate 1 minute on each side.

Break 4: Quad Smash

Whether you're standing or sitting, the fronts of your legs are bound to get tight. To perform this mobilization, you'll need a ball, such as a lacrosse ball or softball (see pages 239 and 240). Simply use your hand to drive the ball into your quadriceps and employ the contract and relax, smash and floss, and pressure wave methods (see "Mobilization Methods" beginning on page 230). The best part of this mobilization is that you can do it while sitting on your stool or chair. Just remember to stand up and move around a little before getting back to work.

If you're at home or you have a private place to mobilize at the office without creeping out your coworkers, you can execute a more effective quad smash; see pages 313 and 314 for details.

Break 5: Shoulder and Chest Opener

This exercise is a great way to improve shoulder mobility and keep your upper body loose during the day. You will need a PVC pipe, wooden dowel, belt, band, or towel—basically anything that measures the length of your arm span and won't rip or break. To begin, take a wide grip on the pipe (or whatever implement you're using). Keeping your arms straight, slowly move the pipe over your head and behind your body. Don't be in a rush when performing this movement. Hang out with your arms behind your body (as in photo 3) to get a nice stretch through your chest and shoulders. After 3 to 5 passes, move your hands a little closer together. Repeat this process until you can't get your hands any closer together.

Break 6: Global Forward Bend

A neutral spinal position is your base working position and the position in which you should spend the most time. But your spine is also designed to bend and twist. Remember, the spine is divided into three segments: cervical, thoracic, and lumbar. Each segment consists of individual bones called vertebrae, which function a lot like hinges. When it comes to stability, these hinges are not meant to bend individually. However, your vertebrae are meant to bend as a part of a global arch through the entire spinal system, which this movement emphasizes. If you do yoga, the Global Forward Bend will look familiar, as there are several variations of the standing and sitting forward bend in yoga.

This bend lengthens your spine and provides a nice stretch through the tissues and muscles of your posterior chain—specifically your back, glutes, and hamstrings. To perform the movement, assume a braced neutral position, take a big breath, and then bend forward starting from your head, exhaling as you bend. As you move toward the floor, allow each vertebra to fold forward in sequence.

So allow your head to fall, then your upper back, and then your lower back. Try to time your breath so that you are exhaling the last of your air as you reach the bottom position. You can grab the backs of your heels and hang out in the bottom position for a few seconds for a deeper stretch. There's no need to rush.

When you're ready, perhaps after a breath or two, go up the same way you came down. But this time, inhale as you come up, pulling yourself into the standing position one vertebra at a time. Be sure to pull your belly button toward your spine as you move. Finish in a neutral position.

We want to emphasize that this is not how you should bend over to pick something up. It is just a nice way to get some motion through your spine.

You can also perform this bend in a chair. It's not as effective because you don't get a stretch through your glutes and hamstrings as you do when standing, but if you're stuck in a chair, you can at least get a stretch through your spine.

Break 7: Glute Smash

Even if you have a standing workstation, you will likely spend some part of your day leaning or sitting on your stool. The Glute Smash will prevent these large muscles from getting stiff. All you need is a ball and a flat surface. Sit on the ball, find an area that feels tight and ropy, and get to work. You can contract and relax (see pages 231 to 233) by flexing your glutes on the ball until the stiffness dissipates, or move your leg around to help "unglue" matted-down tissues.

For a more detailed description of this technique, see pages 299 and 300.

Break 8: Global Rotation

Most people intuitively twist their bodies from side to side when standing for extended periods. Like the Global Forward Bend on pages 172 and 173, Global Rotation is a way to get some motion through your spine. The key to executing this movement is to maintain a neutral position through your pelvis-lumbar complex. If you twist from an overextended position, you are really just hinging on a couple of segments instead of twisting through them all. So keep your belly tight as you turn. Shift your weight onto one leg and come up onto the ball of your opposite foot as you twist. Take it easy with this movement, especially with your initial twists. Rotate back and forth, keeping your arms relaxed, until your back and hips loosen up.

Break 9: Squat

Squatting is one of the easiest and most effective ways to strengthen your lower body and improve your flexibility. The best part is, you can do it right at your desk. There are several ways to approach the squat: you can squat all the way down, squat to a chair, or perform a supported squat. And you're not limited to repetitions. You can and should spend time in the bottom position, especially if your goal is to improve your flexibility. If you're the type of person who needs a benchmark, shoot for 5 to 10 squats, either full range or to a chair, and then spend the remainder of your 2-minute break hanging out in the bottom position, as if you were having dinner in Thailand.

To learn more about squatting technique and other supported squat options, see pages 111 to 115.

Break 10: Anterior Neck Mobilization

When it comes to adopting the forward-head-on-neck position, we're all guilty. It's nearly impossible to keep your head in a neutral alignment 100 percent of the time, especially in front of a computer. As a result, your neck flexors—the muscles in the front of your neck—get short and stiff. The Anterior Neck Mobilization is a great technique for cleaning up that area and preventing tension and pain. The tack and twist method (see page 237) is crucial here: you bind up the tissue by twisting a ball into your neck, and then you look away from the ball and move your head in different directions.

For a full description of this technique, flip to page 258.

Break 11: Arm Circles

Moving your shoulders through the full range of motion is one of the best ways to maintain shoulder flexibility. The keys here are to keep your shoulder in a neutral position as you circle your arm, keep your arm as close to your body as possible, and keep your elbow locked out. Pay close attention to the positioning of Kelly's hand and arm in the photos above. As with all dynamic movements, you want to start out slowly and then steadily increase your speed. Shoot for 10 circles with each arm, then repeat the sequence by circling your arm in the opposite direction. So, if you move your arm backward during the first set, circle your arm forward during the second set.

Break 12: Forearm Smash

If you type a lot or suffer from a repetitive stress injury like carpal tunnel syndrome, put a gold star next to this mobilization. It's a simple, low-tech way to keep the tissues of your forearms supple and healthy. Simply drive a ball into your arm, find a tight spot, and then move your wrist around in all directions. For even better results, rest your arm on your desk and use your other arm to massage the ball into tight areas.

For a more detailed description and some variations of this technique, see pages 292 and 293.

Break 13: The Michael Phelps

We call this the Michael Phelps because swimmers do it as a warm-up before diving in for a race. It's a good way to keep your shoulders loose and get in some upper body movement. To perform this technique, hinge forward from your hips while keeping your belly tight (see pages 106 and 107 for proper hip hinging technique). Spread your arms wide and then hug your chest. As with Arm Circles, you want to keep your shoulders relaxed, start slowly, and increase your speed gradually.

Break 14: Foot Smash

Rolling out your feet on a small ball, like a lacrosse ball, is the easiest way to keep them supple. Your feet are going to get sore and tight, especially if you're standing and moving for most of the day. But it doesn't take much effort to keep them healthy. You can do this technique during a break or while working, so there are no excuses. Place a ball under your foot, apply some pressure, and get to work.

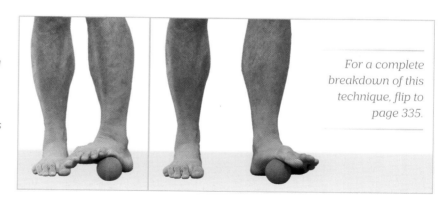

For a complete breakdown of this technique, flip to page 335.

Break 15: Wrist Mobility

This is a nice forearm and wrist stretch that you can do on the ground, on your stool, or in a chair, as demonstrated below. You can perform this mobilization with one hand at a time or with both hands simultaneously. The goal is to accumulate 1 minute with your palm(s) down and 1 minute with your palm(s) up.

Break 16: Shoulder Opener

The Shoulder Opener is a simple way to get some motion and a dynamic stretch through your shoulders. To perform this mobilization, raise your arms overhead and then—keeping your arms straight—throw them back as if you were paddling on a kneeboard. Keep your shoulders relaxed as you swing your arms. It's easy to let your shoulders round forward as your arms swing behind your body, so try to let your shoulders spin in place.

From Sitting to Standing: How to Transition Safely to a Standing Workstation

As standing workstations began to catch on, some critics inaccurately claimed that standing is actually worse for the body than sitting. Given the mountain of research pointing to the fact that sitting too much is horrible for our health, along with the well-documented benefits of standing, these criticisms defy logic. But let's recap what we know.

Sitting is suboptimal for two key reasons. The first is that a sitting body is an inactive one, and, as humans, we are designed to move. Our normal, healthy physiology depends on it. The second reason is that it is nearly impossible to move and maintain a good position while sitting. Standing, on the other hand, creates a movement-rich environment in which you naturally move and change positions and can achieve and maintain good posture.

If you have spent the majority of your life sitting for hours on end, it will take time and effort to transition to a standing workstation. The French social innovator André Godin said, "The quality of our expectations determines the quality of our actions." If you approach your standing desk with the expectation that the transition will take time, you are much more likely to be successful.

For that reason, if you have been sitting at your office for 8 to 10 hours a day, we don't think that you will succeed if you try to spend the same amount of time standing. To make the transition as seamless as possible, you must listen to your body and progress at your own pace.

Once people learn about the negative effects of sitting, they are often over-eager to transition to a healthier way of life. The result? They ditch their sitting workstations and attempt to stand all day, every day. In addition, they don't create an optimal setup or make it a point to incorporate movement and mobility breaks into the workday. This is like taking up jogging to get in shape and then going out and trying to run a marathon after a week. Don't set yourself up for failure. After all, you didn't turn into a sedentary, sitting-adapted person overnight. Adapting to the stimuli of standing, moving, and mobilizing will take time and attention.

While the ultimate goal should certainly be to eliminate optional sitting from your life, you must give your body time to adjust, which may mean spending a portion of your day sitting at the outset. When we first added standing desks to our office, we were shocked at how tired we felt at the end of each day. But, as we practiced standing more and more, that fatigue subsided. After about six weeks, we had transitioned to standing comfortably for the vast majority of the day.

What type of sitting-to-standing ratio should you implement in the beginning? Every person is different, and our best advice is to listen to your body: it is constantly sending you feedback. If your feet and legs start to ache, feed some love to those broken-down tissues by performing some basic maintenance. If that doesn't work, you might need to sit down for 15 or 20 minutes. Our recommendation is to limit your sitting to 20 minutes max—set a timer to keep yourself on track. And try not to allow your body to collapse into the chair.

Remember, too, that not all pain is bad. Sore muscles, for example, are a positive adaptation. Shifting from sitting to standing turns on muscles that may not have been used in years—muscles that take time to develop and strengthen. Learn to distinguish between good pain and bad pain. Making the distinction can be tricky at first, but the good news is that by following the principles outlined in this book, you will become more and more in tune with your body and what it is trying to tell you.

Here are a couple of tips to make the transition from sitting to standing a little easier:

1. **Start slowly.** Rather than trying to stand for the entire day from the get-go, try standing for 20 minutes of every hour you spend at your desk. It doesn't even have to be 20 minutes straight. You can break it up by standing for five minutes and then sitting for 10. If you implement that strategy every hour, you'll have stood for two hours by the end of your eight-hour workday. Not a bad start! This is in addition to walking, moving, and mobilizing every half hour, of course.

2. **Keep progressing.** Stick with your routine of standing for 20 minutes of every hour until your muscles are no longer painfully sore, then progress to 30 minutes of every hour. Depending on your level of health, this might take a week, a month, or longer. The key is to progress at a rate that your body can keep up with. It's not a race, yet you want to constantly push yourself.

If setting a timer isn't your thing, consider standing for certain tasks. For example, if you spend a portion of your day making and receiving calls, commit to standing while you're on the phone. If you spend time answering emails or doing social media work, try standing while performing those jobs. This will get you off your rear end for a manageable amount of time and create a habit that you can stick to.

As long as you keep increasing the amount of time or the number of tasks that you perform while standing, you'll soon be on your feet for the majority of your day. However, if at any time along your journey you feel pain brought on by strain or fatigue, reduce the amount of time you spend standing until your body can catch up.

Optimizing Your Sitting Mechanics

We've been going on and on about the harmful effects of sitting, but of course, some sitting is not optional. Whether you're on a flight, driving to work, or enjoying dinner with your family, sitting is a part of life, and it is here to stay.

Think back to your childhood. Besides your mom or grandma telling you to sit up straight, did anyone ever teach you *how* to sit? We'll assume that the answer is no. But guess what? Sitting is actually a skill that must be learned and cultivated.

As you may recall from Section 2, there are three components, or pillars, of bracing your spine in a neutral position while standing:

1. Engaged glutes

2. Hips screwed into the ground through straight feet

3. Trunk muscles stiffened

The instant you sit down in a chair, it becomes impossible to engage your glutes or screw your hips into the ground, leaving only one of these pillars available: a stiff trunk. In this section, we cover the ways in which you can add pillars of support to your everyday seated positions to make them more ideal.

Let's begin with the way humans are designed to sit: on the ground.

THE THREE GOLDEN RULES OF SITTING

When it comes to sitting, always follow these three rules, regardless of which seated position you adopt:

1. Sit with a neutral spine.

2. Get up and move every 20 to 30 minutes.

3. Perform 10 to 15 minutes of daily body maintenance work.

Sitting on the Ground (Two Pillars)

Sitting on the ground is the most ideal seated position because two pillars are part of the equation: hips and trunk. What's more, when you sit on the ground you automatically sit on your pelvis, which is designed to handle the load of sitting. When you sit in a chair, on the other hand, you typically sit on your hamstrings, which are not load-bearing tissues designed to bear the brunt of sitting. Spend enough time sitting on these delicate wonders of sliding functionality and they'll turn into grilled cheese sandwiches. Want to know how to improve hinging mechanics and hamstring range of motion? Stop sitting on your hamstrings.

Another bonus of sitting on the floor is that, in order to go from standing to the ground, you are forced to squat, which takes your hips through their normal, natural full range of motion. As we noted earlier, cultures that sleep on the ground see far less low back dysfunction, hip disease, and falls in older adults. The ability to get up and down from the floor is also a predictor of overall mortality. A Brazilian study showed that subjects who could not pass a simple test of getting up and down off the floor without support were more likely to die an early death.[1] In other words, getting up and down off the floor is something you must practice regularly. Doing so will literally lengthen your life.

At our house, we sit on the floor while we watch TV. We enforce this rule with our kids, too. This simple rule ensures that we get some ground sitting built into our week. But before you throw away your favorite mid-century modern sofa, know that you can treat your couch like an elevated sitting platform. Just don't let your legs dangle.

Lotus

Sitting on the ground is the best option, but not all seated-on-the-ground positions are created equal. The one that trumps them all is lotus, a cross-legged sitting position in which you place each foot on top of the opposite thigh. Ancient Indian yogis adopted this asana, or posture, to meditate for hours on end, day after day, year after year.

In our model, the lotus position is the most ideal of the seated positions because it offers two out of the three pillars of support. The yogis figured out that this position creates mechanical, passive hip external rotation and stabilizes the pelvis automatically. This is just like screwing your hips into the ground while you are standing. In this cross-legged position, very little trunk tension is needed to create a stable spine. Less work fighting the forces pulling you out of an organized spine means that this position is sustainable.

The problem is, most people can't get into the lotus position because it requires full hip range of motion. And we know that your boss may give you the stink eye if you drop into lotus in the middle of an important client meeting. If you attempt the position and come up short, sit cross-legged (see the following page) and slowly work toward achieving full lotus. And for tips on how to sit during that business meeting, turn to "Sitting Survival," beginning on page 189.

Ever wonder why we are so quick to cross our legs when seated? You guessed it: sitting figure-four (with one ankle placed on top of the opposite thigh) or crossing at the ankles creates an external rotation force. Think of the "social leg cross" as a distant and less successful cousin of the lotus position.

To adopt the lotus position, place your left foot on your right thigh with the sole of your foot facing the sky and your heel close to your abdomen. Next, place your right foot on your left thigh in the same manner. (It doesn't matter which leg is on top; you can switch your feet if you wish.) Both knees should rest on the ground, your torso should be centered above your hips, and your spine should be neutral. While it is important to add some tension to your trunk to keep everything in proper alignment, the tension can be minimal as long as your form is correct. Also make sure to keep from rounding forward into flexion or arching back into overextension.

Cross-Legged

We consider sitting cross-legged to be the second best seated-on-the-ground position. Instead of giving you two pillars of support, sitting cross-legged gives you more like one and a half pillars. You're still able to activate your trunk to brace your spine, but because you don't have as much external rotation in your hips, your pelvis is less stable.

Try this test: sit cross-legged on the floor. If you are unable to adopt this position comfortably, it means that you have lost your normal hip range of motion. Your body has a simple rule: use it or lose it. Having lost it is a symptom of being deskbound.

Sitting cross-legged is an expression of normal physiology. If you can't sit this way, we suggest that you work toward mastering the position. It is the most efficient position for ground-based sitting. It is also easy to practice because it's so versatile; you can sit like this in most office chairs, at restaurants, and even in meetings when your legs are hidden underneath a conference table.

If your hips are tight, sitting cross-legged can force you into a less optimal rounded upper body position. In this case, try sitting on a small pad or hard cushion, like a bolster that you would see in a yoga class. Raising your hips off the floor gives your legs some breathing room and makes it easier for you to sit upright. The key is to position the cushion underneath your pelvis, not your legs. If you're still rounding your back, practice the cross-legged position against a wall. Remember, the best way to improve a position of restriction is to spend time in that position. So, if you're sitting on the ground, try spending a few minutes sitting cross-legged as you cycle between postures.

While sitting cross-legged is a lot easier than sitting in the lotus position, you still need to be mindful of your form. You want to create as much external rotation in your hips as possible by keeping your knees flat on the ground, keep your back straight, and maintain tension in your trunk to stabilize your spine, especially when you are shifting within the position.

Passive Sitting (No Pillars, Really)

Remember the kids' movie *Wall-E*? In case you missed it, it is a dystopian tale in which space-age humans have become obese, infantile consumers who spend their days in hovering lounge chairs, staring at ads on computer screens. This is the image that passive sitting should bring to mind. Passive sitting occurs when zero physical energy is needed to sustain a neutral spinal position because the chair itself is supporting your back, legs, and head. As long as that chair fits your body perfectly, meaning that it doesn't push your head forward or round your back into an odd shape, you don't need to generate any pillars of support because they are built into the chair. The problem is, you can't work from this position—and, of course, you are still sitting.

For a sitting position to be classified as organized and passive, it must adhere to the following guidelines:

1. Your head is in a neutral position.

2. The backrest contours to the natural curve of your spine.

3. The lower back support prevents your pelvis from rotating posteriorly.

4. Your legs and torso form approximately a 135-degree angle.

5. Your legs are either supported or resting on the ground at a 90-degree angle.

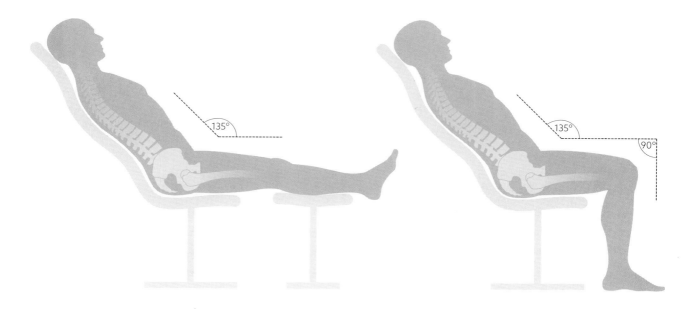

While passive sitting is not well suited to working (unless you are an astronaut launching from Earth), it's a relaxing position to assume at the end of the day. The real trick is finding a recliner that supports your organized passive body rather than contorting your spine and jamming your head forward. Remember, furniture is not one-size-fits-all. You have to find a recliner that fits your body properly.

Maintaining a 135-degree angle between your legs and torso keeps your hips open, which helps prevent your quads and hip flexors from becoming adaptively stiff. It also places very little pressure on your lower back. In fact, according to a study conducted on sitting postures and back pain, leaning back in a chair at about a 135-degree angle is the least challenging angle to your sitting spine.[2] (This is why leaning against a stool at your standing workstation is such a great option—it positions your body at a similar angle.) Ever wonder why your desk chair reclines so far? Or wonder why high school kids look like they are trying to body-bridge in their chairs during class? We humans are clever.

If passive sitting is the best way for you to relax, here are some general tips for buying the right chair for you:

- Choose a recliner with adequate head support.

- Make sure that the length of the seat cushion matches the length of your upper legs.

- Try it out for 30 minutes. If you're following the guidelines in this book, you won't sit for more than 30 minutes without getting up and moving around. After 30 minutes, ask yourself some questions: Does the recliner strain my lower back? Does it support the natural curvature of my spine? Can I maintain a neutral spinal position while sitting in it?

Before you put down this book and run to the recliner store, read the rest of this section. Remember, sitting in a recliner, even in an ergonomically perfect position, is still sitting. For this reason, recliners should be reserved for short periods of relaxation, such as after a long day of standing at your workstation.

Sitting Survival

In most offices, you will rarely find people sitting on the floor or working from recliners. Instead, you will see people sitting upright in their chairs. In this position, one pillar of support is available: the stiff trunk that creates a braced spine. You will also see many people working with their spines shaped like rainbows, reduced to zero pillars of support. Even though most people sit this way for hours on end every day, studies have shown that sitting in a chair is the most challenging position for the spine. Even sitting *well* can put up to 40 percent more pressure on your spine than standing.[3]

Our first sitting survival tip is to create as many pillars of support as you can. It's impossible to activate your glutes while sitting, so your first focus should be on maintaining a stiff trunk/neutral spine—pillar number one.

Then there are some simple ways to add a second pillar of support. One is to sit cross-legged in your chair, as discussed earlier, which adds a rotational element to your hips and stabilizes your pelvis. Another option is to implement what is commonly referred to as the "man spread." It's a gross title, but a fitting one. You simply open your legs wide like a sumo wrestler. Think of "manspreading" (shown on the following page) as crossing your legs without fully bending your knees.

Manspreading

Manspreading creates stability for your pelvis and lower back. You can either position your feet together and let your knees fall to the sides or spread your feet wide. Caution: Manspreading may be NSFW (not safe for work), and you certainly won't make any friends when riding public transportation. Note that while manspreading sounds like a gender-biased activity, it turns out that women also have hips and spines. Wo-manspreading is a thing, too.

Optimizing Your Sitting Environment

As we have already covered:

1. Standing is your best option because it enables you to employ the greatest number of means to support spinal organization and stability. Standing also promotes more movement throughout the day.

2. When standing is not an option, sitting on the ground or in a supported passive position is better than sitting upright in a chair.

3. When you have to sit upright in a chair, prioritize your stiff trunk/ neutral spine and add a second pillar of support by sitting cross-legged or manspreading.

We also fully understand the realities of a busy life. We know that there are situations in which, despite your best intentions, you simply cannot apply any of these strategies. For example, we don't expect you to be able to achieve more than a stiffened trunk while sitting in the middle seat of an airplane or flying an attack helicopter. For these situations, we've created some guidelines to mitigate some of the costs of sitting. Consider these as principles for sitting as correctly as possible in a chair.

1. Get organized while standing.

It is much easier to get your body organized by using the bracing sequence (pages 82 and 83) and *then* sit. Engaging all three pillars of support automatically sets you up for a decent sitting position. By contrast, it is more challenging to sit first and then get organized. When your posture becomes compromised, stand up, go through the bracing sequence again, and then sit back down with your body reorganized. Taking time to reset your position is also a fantastic excuse to get in a little bit of movement.

We've already covered the chair squat, which is how to get into and out of a chair properly. Understanding how to perform this movement is foundational to sitting correctly, so we are including it again here as a refresher. For a more comprehensive description, revisit the "Squatting Mechanics" section on pages 111 to 115.

1. Start by going through the bracing sequence (pages 82 and 83).

2. Keeping your belly tight, drive your hips and hamstrings back and tilt your torso forward.

3. Sit your butt to the edge of the chair so that your weight is distributed over the bony parts of your pelvis (your ischial tuberosities)—see guideline 2.

4. Maintaining about 20 percent trunk tension, sit into an upright, neutral position.

5. When you're ready to stand up, position your feet straight, tilt your torso forward, and get your shins as vertical as possible.

6. Elevate your hips off the chair to load your hips and hamstrings.

7. Stand up by extending your hips and knees.

8. As you stand upright, squeeze your glutes and re-establish a braced neutral position.

Sitting Down

Standing Up

2. Sit on the edge of your seat.

Type the words *office chair* into your search engine, and 95 percent of the chairs that pop up will have armrests, a high backrest, and thick cushions. You may see chairs like the $800 Herman Miller Aeron Chair at the top of the list. Although these works of art were designed with ergonomics in mind, it is still nearly impossible to achieve a good position while seated in these chairs.

Have a look at how your coworkers are using their office chairs. You will likely see them slouched forward with their spines in a C-shape. In order to reach the keyboard or desk, they must reach *way* forward. In this position, their necks are also at strange and unnatural angles.

Sitting all the way back in a cushy office chair puts almost all of your weight on your glutes and hamstrings, which are decidedly not weight-bearing surfaces. Don't believe us? Take a moment and check out the bottoms of your feet. That is what a weight-bearing surface on the human body looks like. The skin is thick and callused, ready to support the weight of your body for a long, long time. Now, picture the most beautiful person in the world. Imagine that butt. Does it look like the bottom of your foot? Of course it doesn't. Butts are not meant to be weight-bearing surfaces, so they should not look like feet.

Deskbound workers often abandon those fancy ergonomic backrests and armrests altogether and lean forward to reach the keyboard or desk. Again, because most people have never been taught how to sit and don't have a schema for understanding how to create a braced spine, they are still sitting with those dreadful C-shaped spines.

This, by the way, is the same position that children adopt in school chairs. It turns out that children's chairs are designed to be stackable so that they are easy for the janitorial staff to move around. As a result, children of vastly different shapes and sizes are forced to sit in chairs that bear no relation to their unique physiologies. Their solution? They sit at the edges of their chairs to reach their desks, and invariably their backs are terribly rounded. No wonder most of us are so comfortable with working postures that would make our grandmothers cringe. We start practicing spinal disrespect all the way back in elementary school.

We recommend ignoring the backrest and armrests entirely and sitting at the edge of your seat, essentially turning the chair into a stool. Sit with your feet planted firmly on the ground, and be mindful of creating a stiff trunk. According to Galen Cranz, professor of architecture at the University of California at Berkeley and author of *The Chair*, you want 60 percent of your weight resting over the bones at the bottom of your pelvis (your ischial tuberosities) and the other 40 percent transferred down to your heels.[4] It should look like this:

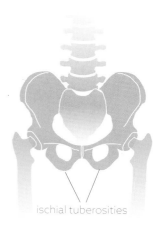

ischial tuberosities

Position the crease of your butt right on the edge of the chair. Your knees should be just outside hip width to generate some hip rotation and stability.

Sitting on the edge of your seat and not utilizing the backrest has two distinct advantages:

- It encourages you to keep your trunk tight, which allows you to maintain a more supported spine.

- It keeps your weight off your femurs and hamstrings. One of the primary mechanisms for hip impingement in adults is chair sitting. Sitting primarily on your pelvis allows your femurs to be femurs. Sitting on your hamstrings, however, drives your femurs up to the tops of your hip sockets. This chronic and abnormal hip joint relationship greatly affects how the femur is able to rotate within the hip socket, pinching the top of the femur and the top of the hip together. You end up with highly compromised hip range of motion.

3. Shift positions as often as possible.

If you have no choice but to sit in a chair for long periods, it is mission critical that you change your position frequently. You probably naturally fidget all day at your desk, but bear in mind that fidgeting is your body's way of telling you that it's time to move. As we have said many times, human beings are designed to be in motion during waking hours. Fidgeting is good, especially for children, but we want you to think of it simply as a cue to move, not as a solution. Sitting-based fidgeting usually results in exchanging one compromised position for another. And many of you have learned to override your body's cues. In our physical therapy clinic, we often see clients who report that they have been feeling symptoms of a problem for months, but it never occurred to them to do something about it. They ignored those cues until their bodies switched warning tactics.

Kelly is no exception. In the 1990s, when he was competing on the U.S. national whitewater canoe team, he paddled for years in compromised positions dictated by the nature of the canoe, including paddling exclusively on one side of the boat. After years of kneeling in these high-stress, high-load positions, he sustained a pretty severe nerve injury in his neck. It turns out that he'd had symptoms in his right arm for nearly nine months before he started to experience any pain. He ignored those overuse symptoms, which were similar to those of carpal tunnel syndrome, and ended up having to treat his now-screaming symptoms with cortisone injections, prednisone, acupuncture, traction, and massage therapy. But it was too late: he had disregarded his body's warning signs, and his professional paddling career was all but over.

Our advice: LISTEN TO YOUR BODY . . . it will tell you when you need to move and when you are adopting suboptimal positions. On the performance side of our strength and conditioning practice, we regularly have to remind our athletes that pain is one of the ways in which our bodies let us know that we are violating our biomechanics.

If you're confined to a chair, realize that you are not limited to one position. When your body sends you the signal to move, stand up and reset your position. Just as you would shift positions and change the orientation of your stool, you can adopt numerous functional positions in a chair (assuming that you have the right kind of chair—turn to page 197 for tips on making your selection). On the following page, we offer some supplemental sitting positions that you can cycle through while sitting in a chair.

4. Stand up and move every 20 to 30 minutes.

This protocol is exactly the same as it is for a standing workstation. Every 20 to 30 minutes, stand up and perform at least two minutes of movement and/or body maintenance work. You can follow the Sample Movement and Mobility Routine on pages 170 to 178, or you can perform one of the prescriptions outlined in Section 7.

Constantly shifting your position while sitting is a great way to get blood flowing into stagnant areas, but it is not a substitute for actual movement. The

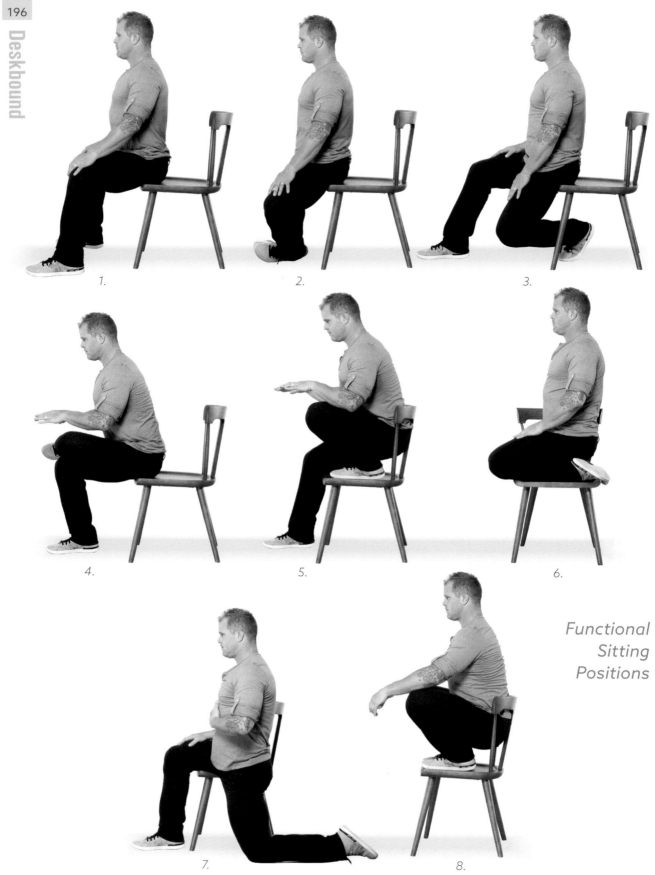

1.

2.

3.

4.

5.

6.

7.

8.

*Functional
Sitting
Positions*

reality is, if you don't get up and move often, you are at risk for lost mobility, weight gain, fatigue, and general ill health.

Keep in mind that regular exercise may not counteract the health hazards of sitting for extended periods, but standing up every 20 minutes certainly will. If you add movement to the equation, like a three-minute walk around your office, you will get even more benefits. According to *New York Times* Phys Ed columnist Gretchen Reynolds, author of the book *The First 20 Minutes,* you will lose weight, lessen your chance of heart disease, and improve your brain function.[5] That sounds like a win-win to us.

If your work environment isn't conducive to taking these two-minute breaks, simply stand up and go through the bracing sequence as often as possible, then sit back down with a neutral spine. Loading your legs, squeezing your butt, opening your hips, and resetting your shoulder position will turn on your musculature and stimulate your body throughout the day. In addition, make it a priority to perform the mobility techniques in Section 7.

The Chair: Choosing the Least of All Evils

Not all office chairs are created equal. Some chairs make it virtually impossible to sit properly, while others provide some flexibility, enabling you to achieve a neutral position. By nature, chairs don't accommodate moving, flexible bodies, so to say that there is an "ideal chair" is an oxymoron, but it's good to know what to look for when making a deal with the devil.

Seat

We prefer to sit in a chair that has a wood or metal seat. If you've ever sat in such a chair, you know that it isn't the most comfortable thing in the world, and that lack of comfort does wonders to promote movement. However, if wood or metal proves too unforgiving for you, you can search for a chair with minimal padding. Just don't go overboard. To find your sweet spot, try sitting in the chair for 5 to 10 minutes before buying it. If your ischial tuberosities start to throb because they're digging into the hard surface, try another chair with slightly more padding. The goal isn't to find a chair that's comfortable to sit in all day long. It needs to be only semi-comfortable for a 20- to 30-minute stretch, because that is as long as you will sit without standing up. If you find yourself in a situation that forces you to be seated, such as working as a driver, pilot, or police officer, treat the padding like we

did the anti-fatigue mat for your feet (see page 143). But the padding should
breathe a little, too.

The shape of the seat should be square instead of round. A square seat
allows you to place the crease of your butt along the edge, as well as easily
find your ischial tuberosities. A rounded seat makes it more difficult to maintain a neutral upright position because your butt will keep sliding off the edge.
It's also important to search for a chair with a wide seat. Ideally, it should be
wide enough that you can sit cross-legged and easily shift your position.

Seat height is another important consideration. Conventional chairs measure 18 inches from floor to seat, which is too tall for most people. As you
now know, you want to sit on the edge of your seat and distribute 40 percent
of your weight onto your feet, which means that both heels must reach the
ground. If the chair is too tall, the entirety of your weight will be placed on
your pelvis. You won't be able to create any additional support from your feet,
which should function like buttresses.

The best way to determine whether the height of a chair is "just right" is to
sit all the way back in the chair and place both feet on the floor. If your heels
lift off the ground and the edge of the chair digs into the backs of your knees
or your hamstrings, keep looking. When you stand next to the chair you're
considering, the top of the seat should be slightly below knee level.

Backrest

The ideal office chair is one that mimics a stool. For this reason, the backrest
is the least important consideration. As proof, look around at the freakishly
expensive "ortho" chairs used by people with a history of back dysfunction.
None of those chairs has a back. Even the chair that James Bond's villainous
foster brother sits in while torturing him in the latest Bond film, *Spectre*, lacks
a back. Now that's evil genius.

Leaning against a backrest that supports your spine is a nice way to relax
and take a load off your body—assuming that the backrest reclines, the chair
adequately supports your head and the natural curvature of your spine, and
you're sitting passively for a short spurt to unwind. If you're actively sitting,
however, meaning that you are trying to do work with an upright, organized
spine, leaning against a backrest will likely deactivate the muscles of your
trunk, causing you to slowly fall into a compromised position. Your body is
most clever; it will save energy any time it can.

Because few chairs come without some type of back support, it's best go
with a material like wood or metal. As long as the surface is hard and ridged,
you can periodically lean against it and still maintain a neutral position. The
key is not to rely on the backrest. Instead, use it to support supplemental
sitting positions, such as when squatting or kneeling in your chair.

*What about physio
balls, BOSU balls, and
kneeling chairs? We do
not consider these to be
great choices. While a
bouncy ball does promote
small movements with
its constant instability,
managing an organized
spine for sustained periods
is nearly impossible.
Anyone on a ball chair
will ultimately default to
an end-range, tissue-
limited shape as fatigue
sets in or concentration
is lost. The unstable
surface accelerates
postural decay and makes
compensatory slouching
or spinal overextension
worse. Try to stand on
a water bed for an hour
and notice what happens.
Add to this the fact that
it is difficult to bear
weight on your pelvis
and not the soft tissues
of your hamstrings. Most
important, sitting on a ball
is still sitting, with all of its
pitfalls.*

*While a kneeling chair
does open up the hips, it
provides limited dynamic
sitting options and
encourages spending
time in an overextended
shape. When it comes to
sitting in a chair, your best
bet is to go with a simple,
rigid chair with a wide,
semi-hard seat.*

Armrests

We recommend seeking an office chair without armrests. While they offer a nice place to rest your elbows and shoulders occasionally, armrests have some inherent problems:

- They often prevent you from scooting your chair underneath your desk, which can position you too far away from your keyboard and cause you to round forward into a position of flexion.

- They can prevent you from sitting cross-legged or adopting many of the other supplemental sitting positions.

If you can use the armrests for their intended purpose—resting your arms—and get close enough to your keyboard without rounding forward, keep them. But if the armrests prevent you from getting close enough to your keyboard and restrict your movement, look for a chair without them.

The Ergonomic Office Chair

The term ergonomic is buried in our office culture. There are ergonomic desks, ergonomic keyboards, ergonomic computer mice . . . the list goes on and on. When we hear the term ergonomic, we typically think of something that will make our lives more comfortable, safe, and efficient. But comfort doesn't always translate to a safe and efficient position, especially when it comes to sitting. The bottom line is that ergonomic chairs will never solve the problem of sitting because their entire goal is to create an artificial support system for your spine. This is a little like asking which cigarette is safest. While this type of support might work for passive sitting, it doesn't translate to a work environment in which you need to spend the vast majority of your time actively sitting.

In fact, most ergonomic chairs create a new set of problems. Their design makes it nearly impossible to sit on the edge of the seat, forcing you to sit on your hamstrings and glutes instead. Worse, the backrest takes your trunk muscles out of the equation.

Don't get us wrong, there are ergonomic tools that can help facilitate a neutral body, such as ergonomic keyboard and mouse setups. But when it comes to chairs, ergonomics is not doing us any favors. Creating different kinds of donuts that are easier to eat doesn't make donuts less of a challenge to your waistline. If the goal is to make your work environment safer and more efficient, then a chair that promotes autonomous sitting is technically the most ergonomic. But the bottom line remains that sitting with your hips locked at a 90-degree angle is unnatural and always will be.

Airplane and Car Seat Survival Guide

Although you want to do your best to reduce optional sitting in your life, there are situations in which sitting is your only option. When you fly or drive, you don't have the luxury of choosing the right kind of seat or standing up and moving whenever you feel like it. This makes travel the most brutal form of chair-based torture. But, like sitting at a desk, there are some things that you can do to minimize the impacts of sitting. In this section we offer some helpful tips for surviving a long flight or road trip.

Airplane Seat Survival Tips

Have you ever sat in a cramped airplane seat and thought to yourself, "This is going to be a pleasant experience"? We haven't either.

Unless you're flying first class, most airplane seats are narrow and C-shaped. In addition to being wildly uncomfortable, they make it nearly impossible to sit in a neutral position. Unable to sit at the edge of your seat, your only option is to lean against the backrest. Here's what typically happens: because the seat is designed to support your back in a C-shaped curve, your pelvis rotates posteriorly underneath your body and your head, shoulders, and upper back are pushed forward. To make matters worse, the seat is not adjustable, so if you happen to be smaller, taller, or wider than the "average person," you're going to get crushed forward. We have observed that anyone over 5'10" ends up getting flexed into a strange position by the nature of the seat. There's no sugarcoating it; it's horrible. But there is hope. Follow these tips for hacking your airplane seat before the seat hacks you:

Tip 1: Use a lumbar support. Sitting in an airplane (or car seat) is the one situation in which using some kind of external support is useful. A small inflatable pad is nice because it is adjustable. But you can use anything, like a pillow or a rolled-up towel or jacket. The key is to position the support in the right spot. Although these pads are called "lumbar supports," you don't necessarily want to position the support at your lumbar spine. All this does is force you into a position of overextension. Instead, position the support at the base of your rib cage. This allows you to lengthen the always-short tissues of your lumbar spine and get into a more neutral position. The other thing you can do is position it low, underneath your belt line. This will lift you off the seat a little bit, putting you in a more autonomous sitting position—similar to sitting on the edge of your seat. Play around with the support and use it to cycle through different positions.

Tip 2: Make friends with the people sitting on either side of you. This can work to your advantage or disadvantage, depending on your predisposition. If you are a bigger person, we recommend making friends with your neighbors so you don't end up in a battle over the armrests. In fact, if you make friends, you can ask your neighbors if it's okay to lift up the armrests (most airplane armrests have a latch underneath). For example, if you're stuck in the middle seat, you can ask the person sitting in the aisle seat to flip up her armrest so that she can get one leg into the aisle. In a nutshell, flipping up the armrests changes the dynamics of your seating arrangement and gives you a little more space to work with. The other good thing about making nice with the people on either side of you is that they won't be as bothered when you stand periodically (Tip 3) and mobilize in your seat (Tip 4).

Tip 3: Open your hips, shift positions, and stand up as often as possible. The one good thing about airplane travel is that you have the option to stand up and walk up and down the aisle. The only caveat is that getting up every half hour and constantly shifting your position may upset the people imprisoned next to you (unless, of course, you've already made nice). At the very minimum, try to straighten your legs, squeeze your butt and thighs as hard as you can, and force your hips into full extension (straight out in front of you) every 20 to 30 minutes. You can do so while in your seat by straightening your legs and thrusting your hips upward or by simply standing up. Turning on your glutes will help reset those femurs and make you feel a whole lot better.

Tip 4: Mobilize in your seat. Always travel with at least one lacrosse ball or, even better, a lacrosse ball and a softball. Security may give you a hard time, but it's worth it. If you're stuck in a seat, you may as well spend some time mobilizing. You'd be surprised at how many techniques you can do while in your seat. You can work on the knots in your forearms, smash into your chest and neck, or stick the ball behind your back or underneath your hamstrings. The bottom line is that you have options.

Tip 5: Use your seat belt to stabilize your pelvis. When you sit all the way back in an airplane seat and you're not actively braced, your lower back rounds and your pelvis rotates posteriorly underneath your body, similar to sitting on a couch or padded office chair. This puts tremendous compressive stresses on the discs of your lumbar spine. To help minimize the pressure on your lower back, try two simple seat belt tricks.

The first is to put your butt all the way back in the seat and try to get your pelvis as neutral as possible, then tighten the seat belt across your hip bones. This locks your pelvis in place and prevents it from rotating underneath your body. Another option is to loosen the seat belt slightly and bridge your hips into the belt. Similar to locking your pelvis in place, this takes some of the load off your lower back so that you can get some momentary relief.

Tip 6: Wear compression socks. We highly recommend that you invest in a pair of compression socks or tights and wear them religiously when you travel. We suggest this simple fix to every professional athlete and team we work with. Compression is an easy way to assist your body's circulatory and lymphatic systems, especially if you can't get up and move every 20 to 30 minutes. The bonus is that compression will help you avoid the dreaded "cankles," or swollen ankles, that so many of us experience during air travel. You can buy a basic pair of compression socks or tights online for around $25.

Tip 7: Stay hydrated. Flying accelerates hypo-hydration, and hypo-hydrated tissues aren't as elastic and don't resist force as well as hydrated tissues do. Imagine a piece of beef jerky—that's what your muscles and tissues look like when they are hypo-hydrated. Hydrated tissues, on the other hand, look like fresh flank steak. Just make sure that you are absorbing the water you drink. We recommend adding a pinch of salt or a basic electrolyte tablet to your water bottle. Your deskbound tissues will thank you.

Car Seat Survival Tips

Car travel is slightly more problematic than airplane travel because you don't have the option to stand up and move around. However, there are ways to game a car seat so that your body doesn't turn into a stiff mess. Follow these tips and—most important—don't do anything that compromises your ability to drive. In other words, don't get yourself or someone else killed while trying to mobilize in your seat.

Tip 1: Change the orientation of your seat. The worst thing you can do is remain in one position. Now, we're not advocating that you lean your seat back like some kind of pimp, but cycling between a couple of different positions can help reset your body and get blood flowing into compressed and stiff tissues. Most vehicles allow you to adjust your seat bottom forward, backward, up, and down. You can tilt the backrest forward and backward, and in some cars you can even adjust the lumbar support. Speaking of which, we definitely recommend using an external lumbar support. The guidelines are exactly the same as when using a lumbar support on a plane—see Tip 1 under "Airplane Seat Survival Tips" on page 200.

🚗 **Tip 2: Periodically squeeze your glutes and point your toes.** When driving, your feet are typically stuck in a slightly flexed position. To prevent your ankles from stiffening around that position, point your toes frequently, either while stopped or when using the cruise control. Also, squeeze your glutes every so often, as your femurs get jammed into the fronts of the sockets. Squeezing your butt helps reset the position of your femurs, so do it whenever you can—even if it's only one butt cheek at a time.

🚗 **Tip 3: Use the steering wheel to keep your shoulders and upper back organized and stable.** One of the great things about driving is that you can use the steering wheel to stabilize your shoulder position and activate the muscles of your upper back. Simply create a little bit of external rotation to stabilize your shoulders. They say that positioning your hands at nine o'clock and three o'clock is safest. Well, it's probably the best grip for your shoulders, too. Creating organized shoulders, as you might remember, has the added benefit of stabilizing your neck and thoracic spine as well. That steering wheel can create another couple of buttresses to support the all-important spine.

🚗 **Tip 4: Move and mobilize during stops.** The key to minimizing the impact of sitting in a car is to stop frequently and mobilize. When you pull into a rest stop or fill up at a gas station, use that time to get some mobility work done. Throw your foot up on the hood or bumper and stretch out your hamstrings, walk around a bit, move your arms . . . no movement is off-limits. It's the same basic approach that you take at work: move and mobilize. You don't need to be a hero and drive seven hours straight. Plan ahead and stop as often as you can without going insane.

Post-Car Ride / Flight Mobility

One of the most important tips for surviving long bouts of required sitting that we can offer is to mobilize for 10 to 15 minutes after you arrive at your destination. We get it; mobilizing is usually the last thing you feel like doing after a long trip. But prioritize it, even when you get in late. Think about it like this: if you go to bed stiff, you're going to wake up feeling even worse.

Performing Basic Maintenance on Your Body

When we started MobilityWOD.com—a website devoted to resolving pain, preventing injury, and optimizing athletic performance—our mission was to equip people with the knowledge to perform basic maintenance on themselves. In Kelly's previous books, *Becoming a Supple Leopard* and *Ready to Run,* he outlined a system for movement and mobility, which includes strategies for treating and resolving pain as well as improving joint and tissue restrictions. If you're familiar with our work, then the information in this section will resonate, as it is based on the same core principles and techniques.

If you are among the uninitiated, you will learn how to address all the components that limit your ability to get into safe and stable positions, as well as how to take a crack at resolving your own pain-based problems. Put simply, you're about to learn a system for resolving pain and improving range of motion. With this knowledge under your belt, you'll have all the tools you need to perform the mobility prescriptions outlined in Section 7.

A Systematic Approach: Mechanics, Lifestyle, and Mobility

As you've probably gleaned by now, the aches and pains brought on by desk work and other forms of sitting can be prevented and eliminated by moving more, moving well, and performing basic maintenance on your body. Having already provided several strategies for increasing activity and moving with proper form, let's dive into self-maintenance.

At this point, you might be wondering what we mean by "self-maintenance," so allow us to clarify. When we say "perform basic maintenance on your body," as we did back in the introduction, we're referring to mobility techniques, or mobilizations, that address stiff muscles, soft tissue and joint restrictions, and poor mechanics. Back pain, neck pain, achy feet, tight hips—these are all symptoms that can be addressed and prevented by using the mobility techniques outlined in Section 7. It's important to note that these mobilizations may look a lot like foam rolling and classic stretching, but they are much more nuanced than aimlessly rolling out your back on a foam roller or bending over to "stretch" your hamstrings.

In order to live pain-free and ensure optimal joint and tissue health, you need take a strategic and systematic approach. If your lower back hurts, for example, you're not going to resolve your symptoms simply by rolling the soft tissues in that area. To truly eliminate this problem, your plan has to include improving everyday spinal mechanics and addressing soft tissue and joint restrictions in the regions upstream and downstream of your lower back, like your hips and thoracic spine. Similarly, if you can't drop into a deep squat because your ankles are too tight, you're not going to make lasting improvement to your ankle range of motion just by "stretching" your calves. Instead, you'll need to spend some time near the limits of your available range of motion and address any soft tissue restrictions that could be contributing to your restricted position.

If you're severely injured or suffering from chronic pain, don't mess around. Enlist the help of a specialist if you haven't done so already. But you should always feel comfortable and empowered to take a crack at solving problems yourself, whether they are pain symptoms or joint or tissue restrictions. A six-minute visit in a doctor's office (the average amount of time a doctor is able to spend with a patient) or a 30-minute physiotherapy appointment just isn't sufficient to address all the components that are contributing to your dysfunction. It is always appropriate to be an educated and proactive partner in your treatment.

More important, you should understand how to perform basic maintenance so that you can prevent problems altogether, well before your tissues have become so adaptively stiff and dysfunctional that you are in pain in the first place. In other words, you don't need to wait for your lower back to hurt or your hips to get tight to know that you're doing something wrong. This is why mobility must be a daily practice. Spending 10 to 15 minutes a day performing self-maintenance will help you stay ahead of potential problems before they become full-blown firestorms.

It's important to emphasize, however, that self-maintenance is just one part of resolving pain and improving range of motion. In order to make lasting change and truly experience the benefits that mobilizing has to offer, you need to manage two additional components: mechanics and lifestyle.

Let's examine each of these components—mechanics, lifestyle, and mobility (tissue health)—separately so you can better understand how to systematically improve range of motion and resolve or, even better, prevent pain and injury.

Mechanics

We've hammered home the importance of prioritizing mechanics as it relates to movement efficiency, injury prevention, and pain resolution. But what does mechanics have to do with mobility? It's simple: the better your mechanics, the less body maintenance you have to do.

Here's an example to help illustrate our point. When Kelly first started lifting, he was prone to overextending his back in the deadlift and the squat. As a result, his lower back and quads would light up, and he would have to spend hours mobilizing his legs and back afterward. The moment he stopped overextending, the problems went away, which meant that he needed to spend a whole lot less time performing self-maintenance. Dialing in your mechanics is like curing a disease without having to treat the symptoms.

What's more, you can't mobilize and resolve all your problems at once. Change takes time. It's an all day, every day endeavor. But if you understand how to move correctly, you can at least mitigate movement errors that compromise tissues and have the potential to cause dysfunction down the road.

Lifestyle

At the beginning of this book, we recommended that you reduce optional sitting in your life. This falls into the lifestyle category. The idea is that eliminating pain and improving your ability to move freely is much more difficult if you don't address environmental and lifestyle stressors. If you wear high heels or other restrictive shoes, for example, it's difficult to improve ankle mobility or ultimately resolve Achilles pain. Put another way, as long as your body has to adapt to dysfunction caused by environmental loads—this is what we mean by "adaptation errors"—managing pain and improving range of motion will be a long, uphill battle. And it is not limited to sitting and shoes. How you eat, drink, sleep, and exercise also plays a huge role.

We like to think of these seemingly disconnected aspects of life as vital components of what is appropriately called a "physical practice." You don't

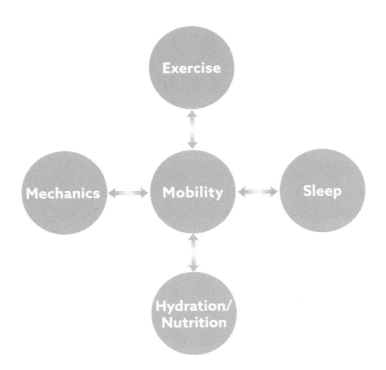

have to exercise seven days a week to have a successful physical practice. But you do need to think about how sleep, stress, and other factors might influence your body's systems. Standing at work, for example, hugely impacts your daily non-exercise-activity demands. We support the idea that small, consistent lifestyle changes can aggregate to significantly improve overall health. You don't need to be heroic; you just need to be consistent.

If you eat sugary, processed food every day, get three hours of sleep every night, and are too busy to breathe hard and maintain your body's mechanics with a movement practice, you can't expect to get completely ahead of potential body dysfunction just by standing at your desk. Even if you follow the guidelines outlined in this section, you still need to improve and optimize the other aspects of your physical lifestyle. For example, jogging and smoking don't really go hand in hand, do they?

Exercise

While exercise may not be able to erase the harmful effects of too much sitting, it is still a critical part of a pain-free, healthy life. As you've probably experienced, organizing and stabilizing your body in a neutral position and then maintaining that position during movement takes work. Well, if your body is weak from a lack of training, then holding these positions is a lot harder. In our experience, people who have a physical practice—whether it's yoga, Pilates, CrossFit, weightlifting, or something else—have less pain and fewer problems than people who don't exercise at all.

To improve your mobility, you need to not only restore your tissues to normal function (see "Mobility Baselines" beginning on page 224 to test yourself), but also be strong in these positions. We think that everyone can benefit from performing basic strength and conditioning movements like squats, deadlifts, and push-ups, but the real key is to dedicate some time each week to developing your physical attributes.

Sleep

Sleep is critical to your physical and mental well-being and your overall quality of life. But you already know this. If you sleep poorly—go to bed late, wake up frequently throughout the night, and get up early—you feel horrible. If we made a list of behaviors that would kill you relatively quickly, not drinking water and not sleeping would be at the top. And don't tell us that you can get by on five hours a night. We do not view this as a badge of honor. The vast majority of adults need seven to nine hours of sleep per night, and the research is clear—even moderate sleep deprivation can have the same effect on the brain as being drunk.[1] Plus, sleep deprivation encourages inactivity, raises your blood sugar, and is linked to poor general health.[2]

There are some simple things you can do to ensure that you get more high-quality sleep. This is known as good sleep hygiene. Yes, temporary bouts of limited sleep are a part of life. Just have a baby and let us know how that goes for you. But even if we aren't always able to control how much sleep we get, we sure can control some of the factors that diminish our sleep quality.

First and foremost, good sleep hygiene means minimizing the use of electronics after dark (see page 157) and sleeping in a cold (between 62 and 68 degrees), pitch-black room. The modern bedroom is like a light show, from glowing clock radio displays to blinking lights from other electronics. Turn them off! And cover those bright-red power standby lights as well. In our house, TVs and cell phones are not allowed in the bedroom, and we cover our old-school alarm clock with a towel to block the light. When we travel, the first thing we do is "sanitize" the room of sleep quality–destroying light sources. If you're not keen on blacking out your window(s) to block the outside light that bleeds in, consider wearing an eye mask and earplugs to ensure a good night's sleep. Your bedroom should be like a closed casket. Sleep like you're dead, right?

If you are among the 70 million Americans who don't sleep well, consider making these simple changes to your sleep hygiene before resorting to an external sleep support like Ambien or NyQuil.

Down-regulation (or getting into a relaxed state) is another key component of good sleep hygiene. We recommend incorporating some deep breathing and soft tissue work into your pre-sleep routine. As we will explain further in a moment, smashing is a basic compression method in which you use a ball or roller to penetrate into deeper layers of muscle tissue. This type of mobility work turns on your parasympathetic nervous system, which signals your body to wind down. (This is why we generally don't recommend doing a lot of smashing before a workout.) Have you ever had a massage? Did you feel like jumping up off the table and sprinting or fighting? No way. Ten to fifteen minutes of soft tissue work (smashing) before bed can have the same effect. You can cherry-pick from the techniques in Section 7—anything that incorporates a ball or roller qualifies. Prescription 2 on page 260 and Prescription 5 on page 278 are good sequences to perform when winding down at night.

This is also a good time to work on diaphragmatic breathing—taking slow, deep breaths (see page 68) as you smash your tissues. It's a two-for-one combo for telling your body to relax.

Hydration

Eating sugary and processed junk is unhealthy. Sometimes fun, maybe, but with sugar being as addictive as heroin, we need to see that stuff for what it is. It's so obvious that it's barely worth mentioning. This is not a diet book, but you should know that for your tissues to function optimally, you need nutrients, which you get from nutrient-dense foods like grass-fed and free-range meats and organic vegetables and fruits. It's that simple. We know too that if you have kids, getting them to eat like real human beings is difficult enough to be a topic for whole other book.

The less obvious component of nutrition is hydration. Being hypo-hydrated (under-hydrated) all the time is a fast track to poor tissue quality and function. The human body is an amazing feat of engineering. There's a lot you can do to keep the engine running optimally, and it all starts with hydration.

One of the keys to improving and maintaining your mobility is ensuring that you have good sliding surfaces, which is how the various tissues of your body—skin, nerves, muscles, and connective tissue (fascia)—interrelate. Do your nerves slide through your muscles? Does your skin slide over your bones and tendons? A chronic state of hypo-hydration can hinder your tissues' ability to slide over one another, limiting range of motion in key areas like your hamstrings and wrists.

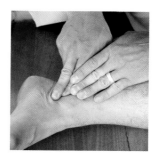

To understand how sliding surfaces work, try this simple test. Sit on the floor or in a chair and cross one leg over the other so that your ankle is resting on your opposite thigh. Now pull up your pant leg (if you have one), exposing your lower leg. Notice that there's a flat part of your shin. Press your thumb or index finger down hard into the flesh over your shinbone near your ankle and hold it for five seconds. Put down this book and give it a try.

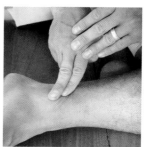

Did you see a dent in your skin? If you did, you have something called "pitting edema," and it's a sign that your tissues are congested and hypo-hydrated. And it's a metric for how well your tissues are sliding over one another. If you're behind on your fluid intake and aren't moving enough, your tissues will become sticky. We call this "tissue lamination." The pitting edema shin test is really an allegory for how all of your tissues are likely to behave.

For example, if you sit on your glutes and hamstrings for an extended period, what do you think is going to happen to the underlying tissues? That's right: butt lamination. All the tissues of your rear end and the backs of your legs are getting sandwiched together under sustained high pressure and heat. And if you fail the hydration test, you don't have any lubrication. You basically turn those beautiful posterior tissues into grilled cheese sandwiches.

So when you're hypo-hydrated, your tissues get sticky. And when they get sticky, you get stiff. When cartilage is properly hydrated, for example, the joints glide. When cartilage is hypo-hydrated, on the other hand, the joints tend not to enjoy the full advantage of their shock-absorbing cartilage.

You can imagine what happens when you freeze your hips and knees into 90-degree positions for 12-plus hours a day.

You need water as well as salt to stay hydrated. But how much? As a baseline, we recommend that you take in a bare minimum of a couple of liters of water a day. All fluid counts toward a well-hydrated state, including the water found in fruits and vegetables (if you eat those) and even the water in tea and coffee.

The "drink more water" message has clearly gotten out, as everyone seems to be carrying around a water bottle these days. The problem is, most people are not actually absorbing all the water they are drinking. Hour after hour, they keep filling up their water bottles, but two strange things happen: their number of trips to the bathroom goes through the roof, and they still feel thirsty.

There is a simple fix for this: add a pinch of salt to your water. Your digestive system will be better able to absorb the water into your tissues, and your trips to the bathroom will become less frequent. If you can't stand drinking salted water, there are companies that make tasty little tablets that you can drop into your water bottle that have the same water-absorbing effect as salt. It doesn't matter which brand you choose.

The protocol is different at meals. Salt your food a little, chase it with pure water, and you're good to go.

The game here is to absorb the water that you are already drinking. We know people who drink a cup of coffee or two for breakfast and don't drink anything else until lunch or beyond. If we told you not to drink anything but a cup of coffee for the next 14 hours, you would tell us we were mad. Yet that's precisely what is going on when you include a sleep cycle on top of that coffee-until-noon plan. We have many high-level athlete friends who start their day by chugging a tall glass of water with lemon and a pinch of salt. When the demands of the day hit, they've already taken care of business.

Mobility (Tissue Health)

Mobility is the third and final component of our system for resolving pain and improving range of motion. The reality is that your tissues and joints will get stiff and restricted if you don't maintain them regularly. And here's the rub: improving your mechanics and optimizing your lifestyle will get you only so far. Moving well and practicing healthy lifestyle habits are critical, but they're only two parts of a three-part system. You still need to mobilize your tissues. Even if you eat a perfect diet, sleep eight hours a night, and move with impeccable form, the demands of daily life will catch up with you. Muscles are going to get stiff. You're probably going to experience some musculoskeletal pain eventually. This is why performing 10 to 15 minutes of daily mobility work is one of the four basic guidelines covered at the beginning of this book.

Remember, the goal is to use mobility as a preventive and restorative practice so that you can stay ahead of problems and keep restrictions and pain from setting in. If you're constantly fighting to resolve pain and you never work to improve your joint range of motion and your ability to move well, something is wrong. You need to address your mechanics and your lifestyle in addition to performing basic maintenance on your body.

When it comes to treating the most common types of pain and dysfunction, addressing some of the contractile functions of your musculature—your ability to engage your muscles—and how well your tissues slide will go a long way. You may have heard these two categories be referred to as "myofascial work." The techniques that we use to perform this kind of work are called "sliding surface mobilizations." When it comes to resolving major range of motion problems, we tend to try to address potential restrictions within the joint capsules and surrounding musculature. We call this kind of mobility work "muscle dynamics mobilizations." To be sure, though, both myofascial and joint capsule/tissue restrictions contribute to lost range of motion, and both can be generators of pain.

Sliding Surface Mobilizations

To reiterate what we introduced under "Hydration" earlier in this chapter, "sliding surfaces" refers to the critical interplay between your skin, nerves, muscles, and connective tissues. Consider it a catchall phrase that describes how the different structures and systems of your body relate to one another. Your tissues—skin, nerves, muscles, and tendons—should all slide and glide over each other: your skin should slide over the underlying layers of bone, tendons, and muscles; your nerves should slide through your muscles; and your tissues should slide around your joints. If you're getting a massage or smashing your glutes with a lacrosse ball, you're working on restoring sliding surfaces. Sliding surface mobilizations are used primarily for resolving and treating pain, but also can have an impact on joint and tissue range of motion.

Sliding Surface Hand Test

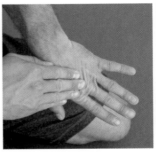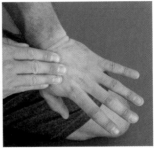

Your skin should slide unrestricted over the back of your hand. To test this idea, take your index and middle fingers, press down on the top of your opposite hand— keeping that hand open—and move your skin around in all directions. See how the skin slides over the underlying bones and tissues? To a lesser degree, this is how your skin ought to slide over all the layers of your muscles, tendons, and bones.

Sliding Surface Mobilization Examples

Quad Smash
(pages 313
and 314)

Glute Smash (pages 299 and 300)

Anterior Neck Mobilization (page 258)

*Triceps Smash
(pages 285 to 287)*

*Hamstring Ball
Smash (page 315)*

Muscle Dynamics Mobilizations

Muscle dynamics techniques can look a lot like traditional stretches and are used primarily for improving range of motion. But don't make the mistake of thinking that these are performed like classic stretches. Most people associate stretching with bringing a joint or tissue to end range and holding it there. The problem with this idea is that it infers that your body's tissues are like rubber bands. They are not. Passively pulling your tissues to end range and then being static neglects the fact that your muscles are run by your brain and nervous system. Yes, you may feel tension, but passive "stretching" is like trying to push-start your car without the key in the ignition. Instead, you want to use an active model—applying tension and slight movement at end range—to help facilitate change and restore tissue function.

Muscle Dynamics–Centric Mobilization Examples

*Single Leg Flexion
and External Rotation
(page 309)*

*Classic
Hamstring
"Stretch"
(pages 318
and 319)*

*Couch Stretch
(pages 307
and 308)*

It's important to realize that a dynamic relationship exists between sliding surface mobilizations and muscle dynamics mobilizations. For example, let's say you mobilize the backs of your legs by smashing your hamstrings (a sliding surface mobilization). You may restore sliding surfaces as well as improve your muscle dynamics, or hamstring range of motion. Similarly, employing a technique that emphasizes muscle dynamics may affect sliding surfaces, too. As you experiment with the techniques in Section 7, you will be employing either a sliding surface mobilization, a muscle dynamics mobilization, or a combination of the two.

Remember, your tissue systems are interconnected and inter-related. You are always addressing multiple systems at once. Certain mobilizations may bias or target a specific system, but stirring a bowl of pasta to move the noodles around is definitely going to affect the surrounding sauce and meatballs.

Now that you understand how mechanics and lifestyle can impact mobility, as well as the underlying approaches to performing self-maintenance (sliding surface mobilizations and muscle dynamics mobilizations), let's apply these approaches to treating pain and improving range of motion.

How to Treat Musculoskeletal Pain

Most of the pain problems that we see fall into the category of "mechanics"-based pain. Less often, pain is caused by catastrophe, like getting hit by a car, or is pathological. Pathology-based pain should be treated very seriously. We always ask our patients and clients if they have unaccounted-for weight loss or gain, changes in bladder or bowel function, or new night sweats, dizziness, fever, vomiting, or nausea. These are symptoms that everyone should know merit a call to the doctor, as that knee pain might be more serious than a symptom of wearing high-heeled shoes.

Musculoskeletal pain (pain affecting bones, muscles, ligaments, tendons, and nerves) is one way for your body to tell you that something is wrong—that you are moving incorrectly, you are in a poor position, you are injured, and/or something is stiff or restricted. As we said earlier, to prevent and resolve musculoskeletal pain, you need to focus on mechanics, lifestyle, and mobility. In this chapter, we will teach you how to attack the symptoms.

It's useful to have some basic guidelines about restoring your body's mechanics. The following "rules" are a great place to start:

1. If something is not in the right place, get it in the right place.

2. If something is not moving, get it moving.

3. Mobilize the area of localized pain.

4. Work upstream (above) and downstream (below) of the problem to address contributing mechanical issues.

To help you understand how to apply these four guidelines, we'll use a basic example. Say the muscles of your upper back are tight, causing minor neck pain. Now let's apply the "rules" so you can see how to treat and resolve the problem.

1. If something is in the wrong place, get it in the right place.

It won't matter how much you mobilize your neck if your thoracic spine is stiffened into a rounded-forward position. Mobilizing your upper spine is a good example of restoring sound joint mechanics first.

2. If something is not moving, get it moving.

One of the easiest ways to eliminate mechanics-based pain is to restore normal range of motion to the problem area. We can't begin to tell you how much pain and dysfunction would be eliminated if our joints could just move through the ranges of motion that they are supposed to. Think of your body as a biological machine. If your car was driving strangely and you got out to look around and noticed that one of your tires was low, you would fix it, even if there were other known issues with your car. Do the same with your body: fix what you can find. Hit upon a stiff tissue? Start there.

3. Mobilize the area of localized pain.

It's important to realize that mobilizing a hotspot or source of pain is not a foolproof plan because it is common to feel pain in one area of your body that actually stems from another area of your body. This is commonly referred to as a "trigger point"—a tight area within muscle or soft tissue that causes pain elsewhere. So, if you have a trigger point in your upper back or trapezius muscle (trap), you might feel pain in your neck and shoulder. That said, it's still useful to mobilize the area of localized pain, especially if you're trying to break a painful muscle spasm. This occurs when a muscle involuntary contracts in a specific area (also referred to as a muscle knot or hotspot).

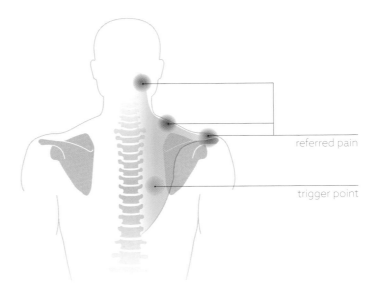

referred pain

trigger point

4. Work upstream (above) and downstream (below) of the problem.

Dr. Ida Rolf—the creator of Structural Integration or "Rolfing"—famously said, "Where you think it is, it ain't." What she's talking about is "referred pain." Your body is encased in a web of fascia—a sheet or band of fibrous connective tissue covering, separating, or binding together your muscles, organs, and soft structures—that transmits movement throughout your body. Tension in one area can impact another, like the ropes holding up a tent. So, if the musculature of your neck and upper back is tight, the fascia surrounding that musculature will also be tight. And if the fascia is tight, it will pull on your cervical spine (neck), potentially causing dysfunction, like pain. By mobilizing the tissues above, below, and around the problem, you can feed slack to the tensioned tissue and restore normal function to the muscles and fascia tugging on your neck. Remember, where the rats get in is not always where they chew.

The upstream-and-downstream approach is great because you really don't need to know anything about movement or anatomy to take care of yourself. Just target the tissues above, below, and around the restricted or painful area. And don't feel limited to just one side; attack your problems from the front and back, too. It can be that simple.

Upstream . . .

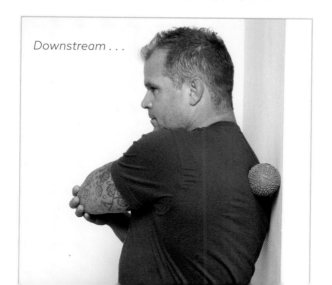

Downstream . . .

How to Improve Range of Motion

Imagine sitting at work or in a car for three hours straight. How do your hips feel when you stand up? They're tight as hell, right? Unable to extend your hips, your body compensates by overextending through the lumbar spine; see the skin-pinch test below. It's a brilliant compensation with horrible ramifications. Your body quickly creates a few extra motion segments by converting your lower back into a couple of extra hip joints.

Skin-Pinch Test

This simple test illuminates how your body compensates to adaptively stiff tissues. Stand up, hinge from your hips, and grab a handful of skin around your hip flexors. Now stand up. What happens? You have to overextend and keep your knees bent to lift your torso upright. This is exactly what happens when you sit for long periods. Your hip flexors start to reflect your working position, becoming adaptively short and stiff.

What you are experiencing when your hips get tight after a long bout of sitting is your body adapting to the position that you've put it in. We call this being "adaptively stiff." You've no doubt experienced it before.

With this knowledge, you can devise a solution. Stand up every 20 to 30 minutes, force your hips through their full range of motion, and employ mobilizations that target the fronts of your hips (see Prescription 9 on page 304). If you're active and you move with good form, your body will adapt to being in constant motion and won't have a chance to stiffen around any single position. This is one way to achieve and maintain full range of motion. The yogis knew this, too. One of the goals of Sun Salutation is to remove the residues of stiffness and tissue change that occurred during sleep. Have you ever seen a cat "stretch" after it wakes up from a nap? Same idea.

Having full range of motion gives you access to a variety of different positions. Imagine how efficient it is to be able to drop into a deep squat to work on the ground without compromising your spinal position or collapsing through your knees and ankles, or to reach your arm overhead without having to compensate by rolling your shoulder forward. When you have full range of motion—the range of motion that you were born with—and you understand the functionally stable positions for your joints and tissues, you greatly reduce the need to compensate (move incorrectly). Humans are brilliant compensating machines. When we eliminate the need to go around our restricted tissue ranges, we tend to revert to our natural mechanics.

But what exactly is full, or normal, range of motion? The simple answer is the ability to get into the basic shapes that average healthy adults should be able to adopt. These shapes aren't crazy gymnastic positions; they represent the bare functionality that every human body should have. On the following pages we offer a few tests that you can use to identify where you are limited and highlight the areas or ranges that you need to work on. If you find that you are missing range of motion in your ankles, for example, you can use Prescription 13 on page 334 to help improve your mobility in that area.

You should be able to get into these positions cold, right out of bed. You should be able to go to the gym, grab a barbell, and start moving in these shapes. We see people who are missing overhead range of motion and they wonder why their shoulders hurt every time they do a pull-up or swim. Well, it's because they are missing overhead range of motion, and that pull-up is forcing a whole bunch of "car accidents" (compensations) in their shoulders. This is why you not only need the range of motion to get into these positions, but also need to know how to get into these positions correctly. You don't need to do pistols (single-leg squats), but you should be able to get into the pistol shape.

Mobility Baselines

Deep Squat Test

The deep (or full) squat is a fundamental human shape that everyone should be able to adopt. Basically, you need to be able to get into this shape to do anything from the ground, whether you're lifting something, gardening, or whatever. Full squatting also means full hip range of motion. Part of managing the sitting compromise is to move your body's structures beyond the limited-range windows required by our modern lifestyle.

The goal is to get your back flat, your feet as straight as possible, and your knees slightly outside your feet. Notice that Kelly's hamstrings are resting on his calves.

If you can't keep your back flat or your butt rotates underneath your body, it's an indication that you're missing hip flexion or ankle range of motion. If you can't get your knees out, it's an indication that you're missing hip rotation and your adductors and groin are likely stiff—see Prescription 9 on page 304 and Prescription 10 on page 312. If you're unable to keep your feet straight, there's a good chance that you're missing ankle dorsiflexion (backward bending of the foot) range of motion and your calves and ankles are restricted—try implementing Prescription 12 on page 326 and Prescription 13 on page 334.

hip external rotation

ankle dorsiflexion

hip flexion

Start by positioning your feet straight—between 0 and 12 degrees—and just outside your shoulders' width. Initiate the movement by loading your hips and hamstrings (see page 96) and lower into a deep squat by bending your knees. As you drop into the squat, try to drive your knees out while keeping your feet in contact with the ground. For a more complete breakdown of this movement, flip back to page 114.

Pistol Test

When you squat with your feet apart, you never take your hips, knees, or ankles quite to their full end ranges. The Deep Squat Test (opposite) assesses hip, knee, and ankle range of motion, but it doesn't capture all the ankle, knee, and hip mobility that you need to be a normal, fully functional human being. This is why the Pistol Test is so important; it measures full hip, knee, and ankle flexion.

There are two ways to perform this test. You can get into the pistol shape (top) or into a deep squat with your feet together (bottom). If you can't get into either of these shapes, it's a clear indication that you're missing hip, knee, and ankle flexion.

This test measures:

· *Hip flexion*

· *Knee flexion*

· *Quadriceps range of motion*

· *Calf and ankle range of motion*

knee flexion

hip flexion

ankle dorsiflexion

knee flexion

hip flexion

ankle dorsiflexion

Starting on the ground— from either a seated position or a deep squat— position your feet next to each other or extend one leg out in front of you, balancing all of your weight on the opposite leg. The goal is to keep your foot straight and your knee tracking outside your foot while keeping your back as flat as possible.

Hip Hinge Test

This test primarily measures posterior chain mobility, specifically in your hip joints, glutes, and hamstrings. Consider this the picking-something-up-off-the-ground shape. The goal is to get your back parallel to the floor with your legs straight. If you can't flex your hips at a 90-degree angle while keeping your shins vertical, you have work to do—see Prescription 8 on page 298.

This test measures:

· *Posterior chain mobility*

· *Glute range of motion*

· *Hamstring range of motion*

Maintaining a braced neutral spine, drive your hamstrings back and tilt your torso forward, allowing your arms to hang. As you hinge from your hips, keep your legs straight and your shins vertical.

posterior chain range of motion

Couch Stretch Test

The Couch Stretch tests hip extension range of motion. It is a wonderful way to assess stiffness through the tissues in the fronts of your thighs. We call it the "Couch Stretch" because we developed it while sitting on the couch watching TV. If this shape is incomplete, adopting and maintaining a neutral spinal position is going to be more difficult. In other words, you need to be able to do it. If you have to arch your back or slide your knee away from the wall, it's fair to say that you're missing hip extension. Being able to get into all the shapes outlined in this section is crucial, but this one is especially important because it highlights the areas that get tight from sitting (the fronts of your hips). Because the Couch Stretch is the opposite of the sitting shape, you might find it ridiculously challenging at first. If you're missing range of motion in this shape, see Prescription 9 on page 304.

This test measures:

· *Hip extension*

· *Hip flexor range of motion*

· *Quadriceps range of motion*

Starting on your hands and knees, position your knee into the corner of a wall, getting your shin flush with the wall. Then, while squeezing the glute on your back leg, lift your torso upright.

hip extension

Overhead Test

The Overhead Test assesses shoulder flexion and external rotation range of motion, which captures any position or movement that requires you to stabilize your arms over your head. If you can't keep your elbows straight, your shoulders round forward, or your elbows flare out, try one of the shoulder prescriptions or the elbow prescription in Section 7. Could your incomplete positioning be caused by stiff triceps, a stiff upper back, or tight lats? Probably.

Keeping your spine neutral, raise your arms overhead. Your arms should be straight up and down. Your elbows should be straight and your shoulders externally rotated. To cue shoulder external rotation, point your thumbs behind you and get your armpits forward.

shoulder flexion and external rotation

This test measures:

· *Shoulder flexion and external rotation*

· *Lat and shoulder range of motion*

· *Triceps range of motion*

Shoulder Internal Rotation Test

This quick test measures shoulder internal rotation. As you can see in the second photo below, the goal is to get your wrist on the same horizontal plane as your shoulder. If your shoulder comes up off the floor as you lower your arm to the ground, it indicates that you're missing shoulder range of motion; try implementing Prescription 3 on page 266. Remember, if you're missing shoulder internal rotation range of motion, your shoulder will roll forward.

This test measures:

· *Shoulder internal rotation*

Lying on the ground, get the back of your shoulder and arm flush with the floor, keeping your forearm vertical. Next, drop your palm and forearm toward the ground, keeping the back of your shoulder in contact with the floor.

shoulder internal rotation

Toe/Foot Test

This test primarily measures toe mobility. You should be able to get your toes to 70 to 90 degrees of flexion. If you're missing range of motion, you're more likely to compensate by walking with your feet turned out. To improve toe and foot mobility, use Prescription 13 on page 334.

This test measures:

· *Toe dorsiflexion*

From a kneeling position, get on your toes and then try to get the backs of your feet at least perpendicular to the ground.

toe
dorsiflexion

Wrist Test

This test measures:

· *Wrist extension*

· *Forearm range of motion*

Stiffness in the wrist joints indicates that the structures of your forearms, wrists, and hands are under more strain than they should be. If you can't get your forearms vertical as demonstrated here, chances are your forearms are restricted. To fix the problem, use Prescription 7 on page 290.

From a kneeling position, rotate your palms away from your body and then place them on the ground with your fingers pointing toward you. The goal is to get your forearms vertical while keeping your palms flush with the floor.

wrist extension

Although the mobilizations outlined in Section 7 will certainly help you improve your range of motion and prevent your joints and tissues from forming a stiffness "cast" around your poor positions, they are not enough to restore or maintain normal range of motion by themselves. Remember, mobility is just one component of a three-part approach. If your mechanics are off and you're making adaptation errors (poor lifestyle choices like sitting too much or wearing heeled shoes), you'll continue to get tight. If there's a hole in a tire, it doesn't matter how much air you put into it; the tire is still leaking. This is analogous to what is happening with your body. Daily mobility work puts air back in the tire, but if you're moving poorly, sitting all day, and not absorbing the water you are drinking, air will continue to leak out of the hole.

Let's say, for example, that you stand with your feet turned out like a duck's, which reinforces movement patterns that trigger calf and ankle stiffness. As long as you're standing and walking like a duck, your ankle mobility will suffer, no matter how much you mobilize. But if you start standing and moving with your feet straight and continue to mobilize, your calves won't get as stiff and your ankle mobility will slowly but surely improve. Our physical therapist friend Gray Cook often says, "Do we move poorly because we get stiff, or do we get stiff because we move poorly?" We think that the answer is both.

In addition to moving well, making optimal lifestyle choices, and committing to daily mobility work, you have to spend time in the positions that you want to improve. As we've said, your body adapts to the positions you put it in throughout the day. So, if you want to improve your squat, you should spend more time in the bottom of the squat. Take the 10-minute squat test: try to hang out in a deep squat or supported squat (see pages 111 to 115) for 10 minutes. If that seems like too much, break it up and do five 2-minute squatting sessions throughout the day.

In other words, to improve your range of motion, you need to spend time mobilizing in the shape that you're trying to change. For example, if you're trying to improve your overhead position, it makes sense to put your arms overhead and mobilize anything that might be limiting your range. Have you ever hung out with gymnasts? They have this annoying habit of walking on their hands. A lot. But they've figured this program out. Being competent overhead and upside down is critical to their success in their sport. Constant "hand standing" is their way of sweeping the cobwebs off of this vital shape.

Identify your position or area of restriction using the mobility baseline tests and then mobilize within the context of the shape or movement you're trying to improve (squat, overhead, etc.).

Also, when you make an improvement in your mobility, you need to *use* that newfound range of motion. You can't expect to restore your tissues to normal function and then not use that "new" range. For example, if you mobilize your hamstrings but never actually hinge or squat, they will likely end up reflecting the old reality that created the dysfunction in the first place. The key is to reflect the change in your movement in an actual movement practice. This is why performing full-range functional movements like squats, deadlifts, and push-ups is so beneficial. In our sports performance practice, we understand the corrective and restorative qualities of movements like squatting—so much so that even in the middle of a brutal world cup competition, we'll have our cycling athletes squat just so that they are able to maintain hip function. Imagine what a few air squats a day will do for you!

Mobilization Methods

On the surface, many of the mobility techniques that we outline in Section 7 look simple. But underlying the photos is a sophisticated system of techniques that help maximize results. We call them mobilization methods. Combining these mobilization methods will give you the most bang for your buck. Turn to page 238 to read about the tools you will use while performing these techniques: a roller, a few different types of balls, and a band.

Say you're working on restoring sliding surfaces to your quadriceps by employing the Quad Smash technique (pages 313 and 314). Aimlessly rolling out your quads is not going to get you the results you're looking for. No, you need to take a more strategic approach. You need to contract and relax to penetrate the deeper layers of muscle tissue, pressure wave across the muscle fibers to undo areas of peak stiffness, and then smash and floss the knotted tissue. The idea is to mix and match in a way that effects the greatest amount of change in the shortest time.

Smash

The smash is a basic compression technique that allows you to penetrate the deeper layers of your muscle and tissue. Just place a ball or roller on the area that you want to improve and then apply pressure, either by distributing your body weight over the tool (as in the Quad Smash, pages 313 and 314) or by driving the tool into your body (for example, by pressing a ball into your forearm for the Forearm Smash, pages 292 and 293).

If you're new to mobility and soft tissue work, expect to feel some discomfort when smashing. The key is to focus on your breathing. When you smash into a sensitive area, the tendency is to contract and hold your breath. Instead, take a deep breath in and then relax as you exhale. This allows your muscles to relax around the mobility tool.

Quad Smash

Forearm Smash

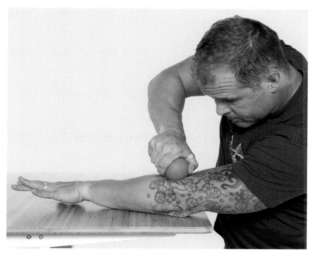

Contract and Relax

Contract and relax is a type of proprioceptive neuromuscular facilitation (PNF) technique. It is widely used for improving range of motion and sliding surface function. It's an active and versatile model of mobilizing and one our favorite mobilization methods. Here's how it works.

If you are trying to improve the range of motion in your legs—for example, by doing the Classic Hamstring "Stretch" shown on the following page—you want to build tension in the position you are trying to change by contracting the muscles, holding for five seconds, and then suddenly releasing the tension. Upon releasing the tension, you will be able to move the leg or joint into a new range. Hold this new range for 10 seconds before repeating the process.

The contract and relax method addresses central nervous system–driven components that may be limiting your muscle function. Building to a strong contraction around the joint that you are trying to improve utilizes features of your neuromuscular system to facilitate change in your contractile tissues (muscles).

Muscle Dynamics Contract and Relax

Contract:
Using a band to keep your foot in place, create tension by driving your leg away from you. Hold that tension for 5 seconds.

Relax:
After 5 seconds, release the tension and pull your foot toward your head until you reach your new end range. Hold that new position for 10 seconds and then repeat the contract and relax process for at least 2 minutes or until you stop making change.

Here's how it typically plays out when you're working on sliding surfaces. Say you're rolling out your quads and you locate a hotspot, an area where you can't distribute your full weight over the roller, or an area where you find it difficult to breathe normally. Your body will naturally tense around that area. One way to get your body to relax and sink deeper into the knotted tissue is to contract the surrounding musculature onto the roller. After five seconds of tension, relax and sink deeper into the tool. Using the contract and relax method in this context not only allows you to go deeper into the problem area, but also provides relief to congested tissue. After a few cycles, you'll notice that the initial tension and involuntary contractions have dissipated. You've provided relief to those tissues. Contract and relax is also the technique that we use to help people reacquire mechanical sensitivity in muscles and tissues that have been injured and are being overly protected by the nervous system.

Sliding Surface Contract and Relax

Contract:
Get as much weight as you can handle over the roller. When you roll over a tight spot, straighten your leg and contract your quads. Hold the contraction for 5 seconds.

Relax:
After 5 seconds, relax your leg and allow the roller to sink deeper into your quads. Repeat this process until you can take your full weight over the roller.

Smash and Floss

Your body is a complex network of interconnected tissue—skin, muscle, tendons, and fascia (to mention a few). When we talk about restoring sliding surfaces, we're talking about dealing with gummed-up, matted-down tissue that can cause dysfunction that could lead to pain or injury. Smash and floss is a crucial method for restoring the natural slide and glide that should exist between layers of tissue.

First you smash, or apply pressure to a hotspot or compromised area. Then you floss by moving your limb around in every direction through as much range of motion as possible ("floss" refers to movement). This method is very similar to active-release treatment (ART) in that it uses pressure in conjunction with movement to de-laminate sticky, matted-down tissues.

After locating a sensitive area or muscle, get as much weight over the ball or roller as you can handle. Then move your limb through as much range of motion as possible. In these examples, Kelly is extending and flexing his leg while keeping his weight distributed over the ball/roller.

Banded Flossing

Sitting in the back of your chair pushes your femurs to the fronts of your hip sockets. This "jamming" creates stiffness within the joint capsules, which are the thick, fibrous bags that hold the heads of your femurs in the sockets.

So, if you sit all day, your femurs move to the fronts of your hip sockets instead of remaining in the center where they should be. Now, every time you perform a deep squat, your femurs hit the fronts of the sockets. "Pinchy" sensations or pain in the front of your hip can be an impingement caused by this poor positioning of the femur within the joint capsule. Using a band to create space or movement within the joint (which we refer to as a "distraction") pulls the head of the femur back to the center where it belongs.

We like to use a jump stretch band for this job. With this thick rubber band, you can actually change how the joint moves within the socket. In more relatable terms, using a band helps reset the position of the joint and its ability to articulate correctly. With the band pulling your joint into a good position, you can create movement through the joint, which we refer to as "flossing."

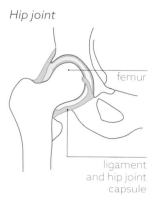

Hip joint

femur

ligament
and hip joint
capsule

In addition to helping reset the position of your femur, using a band helps you manage joint capsule restrictions. When your joint capsules get stiff, it's difficult to effect change using standard mobilizations due to the robust nature of the tissue. To help you understand how this works, imagine a rubber band that is skinny on one end and gets progressively thicker. What do you think will happen when you pull on the rubber band? You'll get more of a stretch from the thinner side. This is one reason you feel more tension in the back of your knee than you do near your butt when you move your leg to end range in a classic hamstring stretch. Your hip capsule is like the thick end of the rubber band. Using a band helps equalize the mobilization forces on these thicker tissue areas.

You can use a band to help clear joint capsule restrictions. When using a band, create a stretch through the band in the opposite direction of the movement you're trying to change. If you're trying to mobilize the fronts of your hips, the band should be pulling you forward. If you're working on your hamstrings, the band should be pulling you backward, as shown.

Pressure Wave

For this method, you place a ball or roller on the target area and then shift your weight over the mobility tool, forcing the tool deeper into the tissue (this is a basic smash). The goal is to relax and sink into the deepest layers of your muscle tissue. Next, you create a pressure wave by slowly rolling over the area you're trying to change, keeping as much weight distributed over the mobility tool as possible. Ideally, you want to work across the grain, or length, of the muscle. For example, in the photos at the top of the next page, Kelly is smashing across the grain of his quads. This is similar to having someone drive an elbow slowly back and forth over a knot. The key is to go slowly and keep the full weight of your body balanced over the ball or roller so that your tissues have a chance to relax into the tool.

The slower you move, the more pressure you will be able to handle and the more benefits your tissues will receive. If you move fast and keep your muscles engaged as you roll, your efforts will be futile. Think of the pressure wave as an icebreaker charging through an ice sheet.

Get as much weight as you can handle over the roller or ball and slowly roll across the grain of the muscle.

Tack and Twist

Tack and twist is another method for improving sliding surface function. Simply press a ball into the tissue you're trying to change and then give the ball a twist. The idea is to take up the soft tissue and skin slack so that you can better target restrictions. Use this technique on areas where circulation is weak and skin gets tacked down to underlying tissues or bone, like the front of your neck near your clavicle and your hands, wrists, elbows, and ankles.

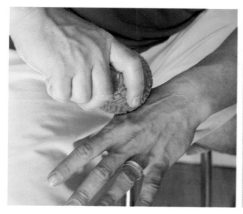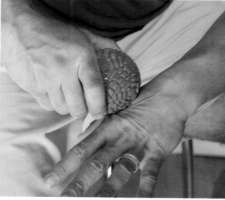

Press the ball into the target area. Then, while maintaining pressure, take up the soft tissue and skin slack by twisting the ball into your flesh.

Mobility Tools

In the introduction to this book, we mentioned that you will need a few pieces of equipment to perform the mobilization methods outlined in the previous chapter. If you haven't picked up these tools yet, don't worry; you can improvise by using whatever you have lying around the house. For example, you can use a wine bottle, water bottle, or rolling pin as a roller. For a small ball you can use a tennis ball, racquetball, or baseball, and for a large ball you can use a softball. The bottom line is to be creative. We never want to hear the excuse, "I don't have the tools, so I can't do it." Use your imagination and make use of what's in reach!

Although anything that you have lying around the house is fair game, some tools are better than others. For this reason, we offer a list of options for each category, most of which you can find online or at your local sporting goods store.

Which mobility tools are the best? That is entirely subjective and depends on your personal preference. Our suggestion is to start with the most basic and inexpensive options and then upgrade from there if you'd like.

As you make your selections, also consider your pain/tissue tolerance. If you enjoy a hard massage (such as a sports, Thai, shiatsu, or deep-tissue massage), we recommend using a harder tool such as a lacrosse ball. If not, we recommend that you start with a tennis ball or softer therapy ball.

Roller

The foam roller is the most common mobility tool. It is great for creating large, dispersed pressures over multiple muscle groups. It's comparable to massaging with a palm or foot in that it has a large surface area for smashing tissues.

If you're new to smashing, a foam roller is probably your best bet. You can get a soft foam roller, which is typically white, or a firm foam roller, which is typically black. In order to effect real change in your tissues, you need to create more pressure and shear than most conventional foam rollers can offer. If you stop feeling change or the foam roller doesn't provide the necessary pressure, consider upgrading to a harder tool, like a wide PVC pipe or the MobilityWOD BattleStar, which has grooves and is specifically engineered for mobility work.

It is a sign of how dysfunctional our tissues really are that something as soft as the ubiquitous white foam roller can cause a pain reaction.

Good for:

smashing multiple muscle groups, contract and relax, pressure wave, smash and floss

Tool options:

foam roller, Battlestar, wine bottle, PVC pipe, barbell

Foam Roller

Little Battlestar and Big Battlestar

www.mobilitywod.com/ product/battle-star

Small Ball

Unlike a roller, a small ball such as lacrosse balls is great for targeting specific hotspots or problem areas. In other words, a small ball is a more precise tool that is used to target smaller areas. If a roller is like a palm, a small ball offers the precision of an elbow. The great thing about a small ball is that you can take it anywhere. It's the perfect companion when you're deskbound.

As with rollers, you have a few options. We typically recommend a lacrosse ball because it is firm enough to sink deep into your tissues and has enough grip to perform a basic tack and twist or smash and floss. The best part is, lacrosse balls are readily available and very inexpensive.

Good for:

precise and acute smashing, contract and relax, pressure wave, tack and twist

Tool options:

lacrosse ball, Supernova 80mm, therapy ball, tennis ball, baseball, racquetball

If you're looking for an upgrade, the Supernova 80mm has a much higher-percentage grip. The teeth of the Supernova are designed to separate the many layers of tissue that need to be penetrated for optimal mobility work.

If you use and prefer a foam roller, we recommend starting out with a small ball that's a little softer, like a therapy ball, racquetball, or tennis ball. *Note:* Tennis balls are designed for tennis. The grip and pliability are not ideal for mobility work, but a tennis ball will get the job done in a pinch or as part of a progression to a firmer ball like a lacrosse ball. Hollow balls tend to collapse, rendering them less effective for mobilizing.

Lacrosse Ball

Supernova 80mm

www.mobilitywod.com/product/supernova-80mm

Tennis Ball

Large Ball

Good for:

targeting trunk musculature and large muscles, contract and relax, pressure wave, smash and floss, tack and twist

Tool options:

Supernova, softball, small medicine ball

If using a small ball is akin to massaging with an elbow, then using a large ball (such as a softball, Supernova, or mini medicine ball) is like massaging with a knee. A large ball has a larger surface area, so the effect is not as acute, but it's great for targeting larger muscle groups like the glutes, hamstrings, quads, and chest. A big ball is also great for smashing the muscles surrounding your torso—specifically your psoas, obliques, and quadratus lumborum (QL).

The most basic option is a softball. The only problem is that most softballs are slick and won't grip your skin. Ideally, you want a ball that has texture or grooves that grip the skin and underlying tissues, like a Supernova. These grooves allow you to tack and twist (page 237) or smash and floss (page 234).

Supernova

www.mobilitywod.com/product/supernova/

Softball

Bouncy Ball

It's a good idea to have a 9-inch bouncy ball, like the ones you can buy from a cage at a toy store. This tool really has only one use: gut smashing (see pages 282 and 283). You can use a kid's ball, a deflated volleyball or soccer ball, or even a small medicine ball.

Good for:
global gut smashing

Tool options:
9-inch child's ball, deflated volleyball or soccer ball

Soccer Ball

Bouncy Ball

Double Ball (Peanut Tool)

The double ball or "peanut tool"—two lacrosse balls taped together with athletic tape—was originally conceptualized for mobilizing the thoracic spine. Working on your upper back with this tool allows you to zero in on a single spinal motion segment or vertebra at a time, which makes for a more focused thoracic mobilization. You can also use a peanut tool to mobilize other areas, like the outer region of your quads, your forearms, and your triceps.

Good for:
mobilizing the thoracic spine, acute smashing, pressure wave

Tool options:
two lacrosse balls or tennis balls taped together, Gemini

As with the other tools, you have a few options. Our favorite is to tape two lacrosse balls together, as demonstrated below.

Step 1

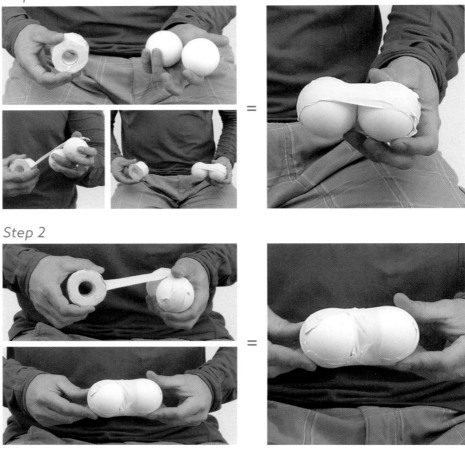

Step 2

The Gemini is another great choice. And if you find the double lacrosse ball or Gemini to be too hard, you can tape two tennis balls together. You can even find dog toys and other mobility products that are similar in shape.

Gemini

*www.mobilitywod.com/
product/gemini*

Band

A simple rubber jump stretch band can be a valuable addition to your mobility tool kit. Using a band is an easy way to address joint capsule restrictions, primarily around the hip and shoulder. You can also use a band to facilitate certain mobilizations. For example, when performing the Classic Hamstring "Stretch" (pages 318 and 319), you can hook your foot through the band and pull your foot closer to your body.

The best Rogue Monster Bands for mobility work are the "light" (green) and "average" (black) size bands. The light band is good for smaller, lighter people, while the average size is better for larger, heavier people. As with rollers and balls, it boils down mostly to personal preference.

Good for:

putting a joint into a good position while mobilizing, facilitating certain mobilizations, contract and relax, banded flossing

Tool options:
Rogue Monster Bands

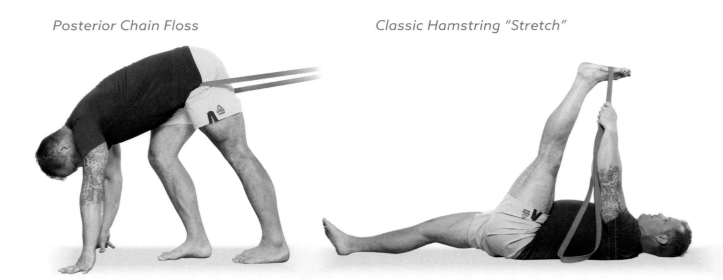

Posterior Chain Floss

Classic Hamstring "Stretch"

Mobility Guidelines

To maximize your time and keep you safe, we've laid out some general guidelines for performing the mobility techniques that make up the 14 prescriptions in Section 7. Consider and digest each guideline before you tackle those techniques.

If It Feels Sketchy, It Is Sketchy

How do you know when tissue is normal? It's simple: if you apply pressure or compress any soft tissue of your body—say, your quads—it shouldn't hurt. In other words, you should be able to distribute the weight of your leg over a ball or roller without feeling any pain. Pain is an indication that your tissues are tight, restricted, stiff, or knotted up. Put simply, if you feel pain while you mobilize, your tissues are not normal. The tension that you are creating around the ball or roller is exaggerating the inherent stiffness in that area. Normal tissues are able to accommodate localized tension forces.

This is a great way to determine which areas of your body need attention. If you're smashing your quads on a roller, for example, chances are only certain sections will hurt. These are the areas that are stiff, tight, or otherwise not normal. When you hit a region of restricted tissue, it hurts, but as soon as you move past it, there is no pain. That's because supple (normal) tissue doesn't elicit a pain response under pressure.

But you must differentiate between the pain or discomfort that you feel when rolling out stiff and tight muscles and the type of pain that indicates injury. It may seem like a fine line, but it isn't. This is where the "If it feels sketchy, it is sketchy" rule comes into play. If you feel like you're injuring yourself or making a painful area worse, you probably are. As mentioned in our book *Becoming a Supple Leopard*, if it feels like tearing, something is probably tearing. If you experience hot, burning pain, your body is telling you that something is not right. If you feel like you're getting a horrible hip impingement, guess what? You're getting a horrible hip impingement. Don't keep mobilizing into the problem, because it will only make the problem worse.

Mobilizing is uncomfortable and often painful. But if you discover a painful area under compression, you stop, and the pain stops, this is a good sign that you are barking up the right tree. Are you holding your breath because it is so uncomfortable? Then you are going too deep. First and foremost, respect your intuition.

We often say, "Don't go into the pain cave." People have an immense capacity to hurt themselves, ignore pain, and travel to places of extreme suffering. Your mobility practice should include none of this stoicism. Stand at the entrance of the pain cave, but do not enter. Mobilizing might be uncomfortable, but it should not be unbearable. Don't be heroic; be consistent in your daily practice. You'll be shocked at how fast your body will normalize. Our bodies are incredible healing and adapting machines. There is no age at which our tissues cannot change for the better.

Stay on an Area Until You Make Change or Stop Making Change

When it comes to mobilizing in a certain position—say, the bottom of the squat—or smashing a tissue such as your quads, the general rule is to stay in that position or in that area for at least two minutes. Consider that your minimum therapeutic dose for making positive change to your tissues or improving range of motion. Kelly's rule as a therapist is to work on a restricted area until there is change or he stops making change. Let's say that he is trying to improve someone's shoulder range of motion. He works on the restricted tissue for at least a couple of minutes. If he's still making change, meaning that he's continuing to improve joint and tissue range of motion or restore suppleness to the tissue, then he keeps going. The moment he stops making change, meaning that he is no longer improving range of motion or reducing stiffness in the tissue, he moves on to the next problem.

What you need to realize is that there is such a thing as over-mobilizing. The chances of your clearing a knot or achieving full tissue and joint range of motion in a 10- to 15-minute mobility session are small. That is why mobility is a lifestyle, an everyday practice. We guarantee that if you lay your quads on a hard ball and go to town for 30 minutes, you will be sore the next day, and possibly the day after.

Here's a more relatable example: Say you're smashing your quads. In the beginning, you'll notice that the tension lessens with every contract and relax and every pressure wave. You can actually feel the muscles relaxing as you employ the mobilization methods outlined earlier. But at a certain point—usually after about two minutes—you reach a point where the tension remains the same. It doesn't matter which method you employ; the state of the tissue doesn't change. The size of the knot stays the same, your range of motion stops improving, and the level of stiffness is the same. That is the time to move on to another area or work upstream or downstream of the area you're trying to improve. To better understand when and how you make change, use the test and retest model.

Test and Retest

Our metric is that if you can't see or feel change, there's been no change. It's very simple. Change should be observable, measurable, and repeatable. If you can't observe or directly measure change, then what you are doing is not working. That's a pretty simple model. So, before you tackle one of the prescriptions in Section 7, consider the movement or position you're trying to improve or the problem you're trying to solve.

Let's say you are trying to improve hip extension range of motion or improve your ability to stand in a braced neutral position. Before you perform the mobility prescription that improves hip range of motion (Prescription 9 on page 304), get into a braced neutral position and check your posture. This is your test. Can you fully extend your hips? Are you able to get your pelvis organized? How hard are you working in this shape? These are the types of questions that you should ask yourself. Now perform the prescription, keeping your intention to improve hip extension as your primary focus. Upon completing the sequence of mobility techniques, recheck, or "retest," your position. Is it easier to stand up straight? Can you get your pelvis into a neutral position? If the answer to these questions is yes, then you know that it worked. You also know or can probably determine which technique(s) had the biggest impact.

Pain is even easier to test and retest. Before you begin a sequence, focus your attention on the area of pain, then perform the prescription or techniques that you think will alleviate your symptoms. Afterward, is the pain better, worse, or the same? If you feel less pain, you know that it worked. If you feel the same, you might need to work on another area or implement additional techniques. If you feel worse, stop mobilizing and seek the help of an expert.

We can't emphasize enough the importance of the test and retest model. It is how you figure out what works and what doesn't. It puts you in a mindset to attack a problem systematically so that you can find the most streamlined solution. It also invites you to be creative when working on your body. Most important, it gives your mobility session context and purpose.

Always Prioritize Mechanics While Mobilizing

Most people have a heightened sense of their position when working out at the gym or playing sports because efficiency is how you maximize performance. It's also the arena in which you're the most susceptible to injury. So, by prioritizing mechanics, you maximize your performance and reduce your potential for injury.

At the beginning of this book, we asked you to apply this same line of thinking to your daily life. The idea is that you should be improving as an athletic human or, at the very least, minimizing the wear-and-tear caused by being deskbound. Well, you should apply this same focus to mobility work. One of the most common mistakes that people make when mobilizing is to ignore the basic body principles (see Section 2). If you're mobilizing with your back overextended or your shoulders rounded forward, for example, all you're doing is ingraining poor mechanics.

Always prioritize mechanics while mobilizing. Just as when sitting, standing, or moving, if you catch yourself in a bad position, stop what you're doing, fix your shape, and then begin again in a better position. This will give you the results you're looking for.

Be Dynamic and Creative

In Section 7, we offer detailed descriptions of how to perform each mobility technique, as well as the area(s) that you want to target when performing each technique. But even though we show you how to perform the techniques, you should treat our instructions as guidelines. In other words, you don't have to perform the mobilization or mobilize the area exactly as demonstrated. In most cases, especially when it comes to the photos, we're showing only one area or a couple of motions or options that you can apply to that particular mobilization.

In other words, you are not limited to performing the techniques exactly as demonstrated. Nobody understands where you are weak, restricted, and stiff better than you do. As long as you prioritize mechanics and employ the mobilization methods, you can target whichever area needs work or move into your own areas of restriction. We call this "informed freestyle." You're using a principle-based approach (organized body, methods, etc.) and targeting areas that you feel are restricted. This means that when you're performing a mobilization, it might look different than what's shown in the photos, and that's okay. Be dynamic and creative. Explore and own your business.

Programming for Mobility

Whole-Body Mobility
Prescriptions

Mobility
Prescriptions

When it comes to resolving pain or injury to joints, muscles, tendons, ligaments, bones, or nerves and improving range of motion, there is no single silver bullet technique. Rather, improving your ability to move safely and effectively requires a systems-based approach. Now that we have given you an overview of our mobility system, the next step is to put it all together into manageable sequences. We call them prescriptions.

In this section, we outline 13 mobility prescriptions covering every region of the body, plus a catchall daily prescription especially tailored to the deskbound crowd. Each prescription comprises three or four mobility techniques and takes between 8 and 18 minutes to complete. Consider these 14 prescriptions a primer for performing daily maintenance on your body.

When approaching these prescriptions, you may be wondering, "When is the best time to mobilize?" Well, the generic answer is anytime you are in pain, feel tight or restricted, or have a window of time to put in some work. But it can be a little more nuanced than that, depending on what you're trying to accomplish. Here are some tips to guide your mobility work and help you make the most of your time:

Starting the day—This is a good time to get your body moving through its full ranges of motion (see "Mobility Baselines" on pages 224 to 228), so prioritize muscle dynamics mobilizations. If you wake up with musculoskeletal pain, some light smashing probably won't hurt.

Throughout the day—Burying mobility techniques into your day is one of the best ways to stay ahead of problems. You don't need to perform a full prescription; one technique at a time is fine. Work on sliding surfaces or muscle dynamics. It's all fair game.

Warming up before exercise—The goal here is to get your body hot. Prepare your body for your sport or exercise with dynamic movements. If you're going to do some heavy squatting, for example, warm up with some light squatlike movements. Avoid heavy smashing and end range stretching before training; it can compromise your mechanics and increase the potential for injury.

Cooling down after exercise—This is the best time to work on improving range of motion. You're hot and your joints, muscles, and tissues are prepped for some quality end range mobilizations. Prioritizing muscle dynamics mobilizations is your best bet here.

Winding down before bed—Sliding surface mobilizations help put you into a relaxed state. Any kind of smashing is great whenever you want to relax.

Programming for Mobility

Taking on a mobility practice is one of the best things that you can do for your body. But, just like learning how to organize and stabilize your body in a good position (see Section 2) and increasing your activity level throughout the day, there's a bit of a learning curve. You have to create and instill a new habit. What's more, it takes time to figure out which mobility techniques will work best for you and how to put those techniques together into personalized sequences—what we call "programming." One person's daily routine probably won't look like someone else's routine. We all have unique forms of stiffness and restriction that require specific techniques and tools. So, before we dive into the prescriptions, let's review some basic guiding principles for mobility programming.

Although we provide specific mobility prescriptions in the pages to come, they are just sample "recipes" to get you heading in the right direction. Ultimately, your goal is to be able to construct your own individualized mobility program based on your particular needs, demands, and problems of the day. That way, you can resolve issues on the spot as they arise. If you are trapped on a long car ride, for example, you'll know exactly what you need to do to minimize the effects. Or, if you tweak your back bending over, you can choose mobilizations that will help that area. Put simply, don't limit yourself to the prescriptions outlined in this section. Find the techniques that work best for you and develop sequences that fit your particular brand of stiffness and restriction.

That said, we always get the question, "What are the top mobilizations that I should do every day?"

The problem with creating a concrete prescription or program is that it results in blind spots. For example, if you do the same four mobilizations every day, you will end up prioritizing some areas and tissues and forgetting about others.

We can say unequivocally that everyone will benefit from doing something that looks like the Couch Stretch (pages 307 and 308), because just about everyone sits and has incomplete hip function. We could probably say the same thing about the T-Spine Smash (page 261), which combats upper back rounding. See Prescription 14, "Deskbound Rx," on page 340 for a sample prescription that is specific to desk work and sitting. But we want you to treat this for what it is: a sample prescription. The problem with set programs is that you get stuck doing the same thing over and over, and you never work on your overhead position, or shoulder internal rotation, or ankle range of motion. And there are so many techniques that attack the same area, all of which have a slightly different effect. For example, one day you might work on improving your sliding surfaces, and the next day you might work on trigger points. As with your diet, variety is the key to success and ensures that you are getting all the essential "nutrients" that you need.

After you go through the 14-day full-body mobility overhaul, which consists of the 14 prescriptions outlined in this section, we suggest that you use these general guidelines to direct your ongoing mobility work:

1. **Attack problem joints and tissues first, and then go after positions of restriction.** Picture a dartboard. Your painful joints and tissues lie at the center, and you should spend a healthy percentage of your 10- to 15-minute session going after that bull's-eye. Spend the rest of your time on improving restricted positions, like squatting.

2. **Write down a list of your problems that could potentially be solved or treated by using mobilization techniques—for example, "low back pain" or "tight hips from sitting."** You'd be surprised at how easy it is to forget what you need to work on when you dive into a mobility session, especially if you have other things on your mind. Think of it like a grocery list–it will ensure that you don't miss anything. Before you start your mobilizations, read over your list and make sure that you understand the importance of what you're about to do. For example, if one of the items on your list is low back pain, think about what it would mean to resolve it. How would your life improve if you no longer had to suffer every time you bent over? This exercise attaches meaning and value to your mobility work. In other words, understanding *why* you're doing something–whether you're trying to improve a position of restriction or resolve pain–gives your mobility session purpose.

3. **Spend no less than two minutes in each position.** Our clinical experience has led us to believe that most people won't hold a position or mobilize long enough to actually make change in the tissue on which they're working. We have found that two minutes is the minimum therapeutic dose per position. If you're doing the Couch Stretch (pages 307 and 308),

for example, spend at least two minutes mobilizing each hip. But it's worth repeating that Kelly's rule as a therapist is to work on a restricted area until there is improvement or he realizes that there's no more improvement to be had in that session. This could mean two minutes or 10 minutes. So don't be in a rush to move on if you haven't experienced positive change in the tissue.

4. **Choose three or four mobilization techniques or areas/positions to mobilize.** Don't get overly ambitious and try to mobilize in 10 different positions. The goal is quality, not quantity. Make some real change in your body. You can fix the rest of it tomorrow.

Remember, programming for mobility changes from day to day, depending on your current areas of restriction.

14-Day Whole-Body Mobility Overhaul

Now that you have a basic understanding of how to program for mobility, let's discuss how to effectively use the mobility prescriptions provided. Again, these prescriptions are just sample sequences. You don't need to follow them exactly as demonstrated. You can perform the prescriptions in any order, or cherry-pick mobilizations from different prescriptions to form your own personalized sequences.

If you're new to mobility work, though, we recommend that you take a two-week challenge and perform every one of the prescriptions. This will not only expose you to all the mobility techniques, but also ensure that you mobilize every region of your body. Everyone has unique problems that contribute to joint and tissue restrictions, poor movement mechanics, and pain. To ensure that all of your body parts are maintained, it's important to cycle through all the areas of your body over the course of two weeks.

After you've performed all 14 of the mobility prescriptions, start experimenting. Mix and match different mobility techniques to form your own personalized prescription based on your unique pain and restrictions. To streamline this process, keep a log or journal of your progress so you know which mobilizations work best for you. In addition to designing your own customized mobility prescriptions, you can continue to use the sample prescriptions provided to attack range of motion limitations, movement and positional faults, and pain.

Using the Mobility Prescriptions to Resolve Pain

If you have a specific pain that you want to resolve, simply go the area that is giving you problems and perform the related prescription. For example, if you're suffering from low back pain, perform Prescription 5: Low Back and Trunk. But remember, resolving pain requires a four-part approach—see page 217. When it comes to mobility work, you need to work on the area of localized pain, as well as mobilize the tissues above, below, and around that area. So, if your low back hurts, you should perform Prescription 9: Hip and/or 10: Upper Leg in addition to Prescription 5. To help you navigate the prescriptions, we've created an area map of the body regions, which you can find on the next page.

Again, after you've gone through all the prescriptions, you will have a better understanding of which techniques bring you the most benefits. To target your personal issues, you might choose to do two techniques from Prescription 6, one from Prescription 7, and another from Prescription 5. There is no wrong way to approach mobility work.

Using the Mobility Prescriptions to Improve Range of Motion

The same approach can be applied to range of motion restrictions. In Section 6, we outlined some quick tests that you can use to determine where you're lacking range of motion in any of the baseline positions (see pages 224 to 229). If you took these tests and realized that your range of motion is limited or you want to improve a particular position, you can use the list of prescriptions on page 255 to help you select the prescription(s) that will improve that position, in order of importance. For example, if you want to improve the squat shape, simply perform Prescriptions 9, 8, and 10, with Prescription 9 being the most critical. Just as when using the prescriptions to resolve pain, the goal is to select the techniques that will have the biggest impact on your mobility. And remember, the best way to improve range of motion in a position of restriction is to mobilize in that shape. So, if you're trying to improve your squat position, mobilize in positions that are similar to the squat shape.

Whole-Body Mobility Prescriptions

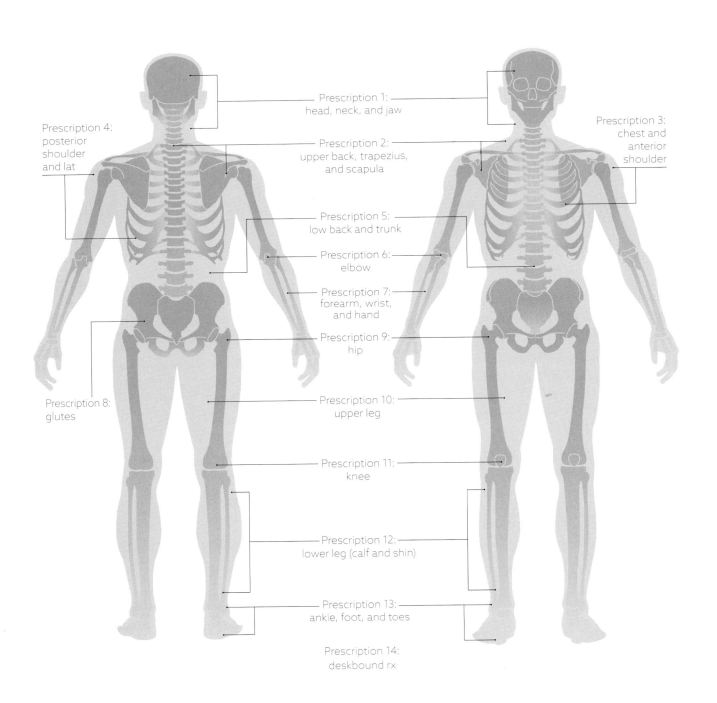

Prescription 1:
head, neck, and jaw

Prescription 3:
chest and anterior shoulder

Prescription 4:
posterior shoulder and lat

Prescription 2:
upper back, trapezius, and scapula

Prescription 5:
low back and trunk

Prescription 6:
elbow

Prescription 7:
forearm, wrist, and hand

Prescription 9:
hip

Prescription 8:
glutes

Prescription 10:
upper leg

Prescription 11:
knee

Prescription 12:
lower leg (calf and shin)

Prescription 13:
ankle, foot, and toes

Prescription 14:
deskbound rx

Pain Symptoms

TMJD	Prescription: 1
Headache	Prescriptions: 1, 2
Neck pain	Prescriptions: 1, 2
Upper trap tweak/pain	Prescriptions: 2, 4
Upper/middle back pain	Prescription: 2
Shoulder pain	Prescriptions: 2, 3, 4
Anterior shoulder pain	Prescription: 3
Posterior shoulder pain (rotator cuff problems)	Prescription: 4
Low back pain	Prescriptions: 5, 8, 9, 10
Elbow pain (tennis elbow, golfer's elbow)	Prescriptions: 6, 7
Carpal tunnel syndrome	Prescriptions: 7, 6
Wrist pain	Prescription: 7
Thumb and hand pain	Prescription: 7
Hip pain	Prescriptions: 9, 8, 5
Sciatica	Prescriptions: 8, 10
Knee pain	Prescriptions: 11, 10, 12
Shin splints	Prescription: 12
Ankle/Achilles pain	Prescriptions: 13, 12
Plantar fasciitis (foot pain)	Prescriptions: 13, 12
Bunion/turf toe (toe pain)	Prescription: 13

Range of Motion (Mobility Baselines)

Squat	Prescriptions: 9, 8, 10
Pistol	Prescriptions: 13, 11
Hip bend	Prescriptions: 10, 8
Couch stretch (hip extension)	Prescription: 9
Overhead	Prescriptions: 4, 2
Shoulder internal rotation	Prescriptions: 3, 4, 2
Toe/foot	Prescription: 13
Wrist	Prescription: 7

PRESCRIPTION 1
Head, Neck, and Jaw

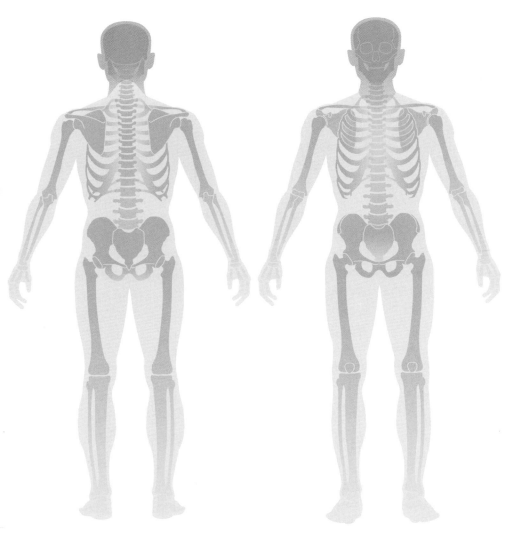

This prescription can be used to treat these symptoms and restrictions:

- Headache
- Neck pain and restrictions
- Tension from forward-head-on-neck position fault
- TMJD (jaw-related pain)

Methods:
- Contract and relax
- Pressure wave
- Smash and floss
- Tack and twist

Tools:
- Small ball
- Large ball

Total time:
8 to 14 minutes

Overview

If you stare at a glowing screen for a living, you grind your teeth at night, or you suffer from headaches, this prescription is a great place to start. Consider it a catchall mobility prescription for tension headaches and jaw pain.

Remember, you're responsible for mobilizing everything under your hair-line. This includes your head, face, and jaw. Think about the number of duty cycles that you put your jaw through. Chewing, grinding, talking, clenching, and mouth breathing can take their toll. Anyone who suffers from temporo-mandibular joint disorder (TMJD) will tell you that they would rather cut off a toe than deal with the pain. It's serious, and it's a hard bell to un-ring. This prescription can both alleviate that pain and put you on the road to recovery. (If you grind your teeth at night, we highly recommend that you see your dentist and get fitted for a nighttime grinding guard.)

Your posture also plays a big role in creating jaw and neck–related mechanical dysfunction. Positioning your head out in front of your body not only strains your neck, but also creates a challenging position for your jaw joint. We're all guilty of positioning our heads out in front of our bodies. It's an easy postural default, especially when you're lurching forward to look at a screen or phone. Fortunately, it's easy to correct and treat. Fix your position by resetting your head and shoulders—see "The Bracing Sequence" on pages 82 and 83—and then use this prescription to treat the symptoms.

Head Mobilization

When it comes to mobilizing, we don't generally think of the head as being an area that needs to be addressed. But it is essential, especially for those of us who are deskbound. This easy mobilization helps alleviate common tension headache symptoms. The best part is, it can be done at your desk. So grab a ball, wind it up in your temple, and then make as many funny faces as possible—roll your eyes, raise your eyebrows, open and close your mouth, etc. If anyone is watching, you might get some funny looks because, well, you probably look funny.

1 to 2 minutes on each side

1) Position a small ball on your temple, next to your eye, and apply pressure.

2) To take up the soft tissue slack in your face, spin the ball into your head.

3) Pressing the ball into your head, open and close your mouth, raise your eyebrows, and make circles with your eyes.

Deskbound

Jaw Mobilization

1 to 2 minutes on each side

The temporomandibular joint connects your jawbone to your skull. When this area gets stiff and moves poorly, TMJD is a potential result. Keeping the muscles of your face and jaw supple is paramount to managing jaw-related problems. Remember, you want to clear the tissues and mechanics above and below the problem area. So, if your jaw is giving you grief, you need to focus not only on the creaky jaw joint, but also on your temples and neck flexors.

1) Position a ball on your jaw joint, or on the muscles of your jaw right in front of your ear.

2) Maintaining steady pressure, twist your hand. Move your jaw up and down and side to side.

3) You can also clench your teeth to contract and then relax, pressing the ball deeper into the muscle.

Anterior Neck Mobilization

1 to 2 minutes on each side

Slouching puts the neck flexor muscles at the front of your neck, from your jaw to the outline of your collarbone, on tension, which can cause them to become stiff and painful. Tightness in this area can cause problems galore: neck pain, headaches, and compromised jaw mechanics (to mention a few possibilities). This area of your neck should feel like soft butter, not beef jerky. So get in there with a ball and restore sliding surfaces and range of motion to those gristly tissues.

1) Tilt your head to the side, pin a ball to the side of your neck, and apply pressure.

2) While pressing the ball into your neck, bind up the tissue by twisting the ball in place.

3) Tilt your head away from the ball and move your head in different directions. Remember, the target area runs from your jaw down to your collarbone and around the front of your neck.

Posterior Neck Mobilization

Stabilizing your 10-pound head on your shoulders is no joke. Your neck extensors—the muscles running down the back of your neck—have a big job to do. Slouching makes these muscles work overtime, so give them some love. Get a large ball like a Supernova and scour the outline of your skull. When you find a tight spot, scrub back and forth across the knotted muscles by rotating your head back and forth. Good riddance, headaches and neck pain.

2 minutes

1) Lying on your back, position a large ball in the center of your skull. You will feel a small pocket. Place one hand on your forehead to add pressure.

2) Slowly rotate your head to the side, rolling the ball along the base of your skull to your ear.

3) Turn your head to the opposite side, scouring the entire base of your skull.

4) To add even more pressure, cross your other arm over your forehead, grabbing the front of your forearm with your bottom hand.

PRESCRIPTION 2
Upper Back, Trapezius, and Scapula

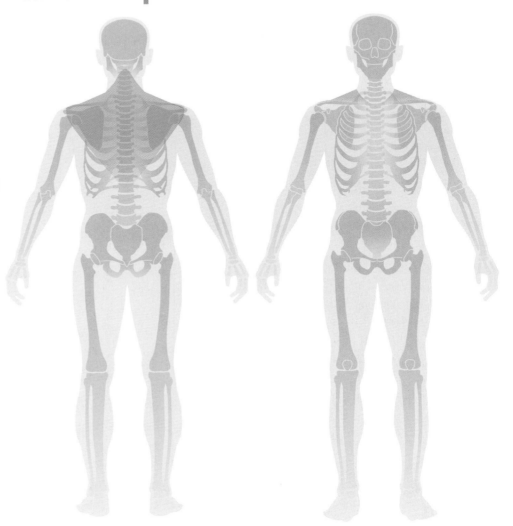

This prescription can be used to treat these symptoms and restrictions:

- Neck pain and restrictions
- Overhead range of motion
- Tension from forward-head-on-neck position fault
- Tension headache
- Thoracic (upper back) restrictions and stiffness
- Upper trapezius tweak/pain

Methods:
- Contract and relax
- Pressure wave
- Smash and floss

Tools:
- Roller
- Peanut tool
- Small ball

Total time:
10 minutes

Overview

The forward-head-on-neck position can wreak havoc on your neck and shoulder function. It can cause neck pain, headaches, muscle tension, and loss of crucial range of motion, like when you are looking over your shoulder to back up your car.

The cure is twofold. First, you need to improve the relationship between your head and trunk positioning. Second, you need to improve the quality of the compromised tissue system that has been working so hard to manage your inefficient mechanics.

This prescription is not limited to correcting and resolving pain associated with the forward-head-on-neck position, though. Use these techniques to treat headaches or tension or pain in your upper back, neck, and shoulders.

2 minutes

T-Spine Smash

When your thoracic spine gets stiff, it becomes quite difficult to effectively organize and stabilize your shoulders and head. A stiff thoracic spine can also result in neck and shoulder dysfunction. The T-Spine Smash is our go-to technique for improving thoracic (upper and middle back) mobility. The key to performing this mobilization correctly is to create a teeter-totter effect over the area that feels stiff. So find where your back is tight by rolling up and down your upper and middle back. When you find a tight spot, use the mobility tool as a fulcrum by arching your back over it. You can also roll from side to side and seesaw through some of the soft tissue adjacent to your spine, or raise your arms overhead to tie in your shoulders. There is no set sequence; the idea is to combine positions. Start from the hug position, roll from side to side, raise your arms overhead, and then go back to the hug. Find where your back is tight and then stay on that area—breathing deeply, arching, and moving. Stay in one place until you feel your tissue quality improve or your symptoms change.

You can use a roller or a peanut tool for this mobilization. A roller is great for targeting multiple spinal segments, while a peanut tool (shown on page 263) allows you to target individual spinal segments.

1) Sit on the floor and position a roller on your back at the base of your rib cage.

2) Wrap your arms over your chest in a big hug. This takes up the soft tissue slack in your upper back and pulls the scapulae out of the way, allowing you to target your thoracic spine.

3) Keeping the roller in the same spot, arch back over the roller. From this position, take a big breath and try to arch farther over the roller as you exhale. You can also elevate your hips to increase the pressure.

T-Spine Smash: Side-to-Side Variation

T-Spine Smash: Overhead Variation

4) With your arms wrapped tightly around your body, sit up as if you were doing a crunch.

7) Return to the start position.

5) Keeping the majority of your body weight over the roller, twist from your hips or rotate your entire upper body. From here, you can arch or side-bend over the roller or roll up and down the side of your back.

8) With your abs braced, raise your arms overhead, interlocking your thumbs. Keep your elbows locked out and reach for the ceiling.

6) You can also roll back and forth, seesawing from side to side.

9) Arch back over the roller.

10) To add more pressure, drive your heels into the ground and elevate your hips.

Note: You can combine the arch with the hip lift as demonstrated here, or isolate the technique. In other words, you can arch, sit up, and repeat, or you can arch, elevate your hips, drop your hips, and repeat.

11) Still arching back, lower your butt to the floor.

T-Spine Smash: Peanut Tool Variation

Using a roller to mobilize your thoracic spine is a great way to open up your entire upper back. But, as we mentioned, it's difficult to target specific areas of stiffness. For a more targeted approach, use a peanut tool to home in on the restricted vertebrae or region giving you trouble. The peanut tool is particularly good for targeting your cervical spine, around your traps and shoulders. As with a roller, you can arch your back, elevate your hips and lower them to the ground while arching back, rotate from side to side, raise your arms overhead, or employ a combination of these movements.

1) Hug your arms around your body to take up the soft tissue slack and move your scapulae out of the way.

2) Try to get the points of the peanut tool between your vertebral segments or spinal discs.

3) When targeting your traps, shoulders, and neck, elevate your hips to increase the pressure and grab the back of your head.

4) Keeping your hips elevated, slowly bring your chin to your chest and sit up as if you were doing a crunch.

Trap and Shoulder Smash

Your trapezius muscles (traps) are responsible for stabilizing and moving your shoulders and neck. Stiffness in this region can cause a lot of pain and mechanical problems around your neck and shoulders. And if you're like most people, you carry a lot of tension in your upper back. The Trap and Shoulder Smash is a great way to restore suppleness to this area.

The first time you mobilize this area, it probably won't take you long to find a hotspot (sensitive area). The key is to get as much weight as you can handle over the ball and then floss by moving your arm around in all directions. Your traps are robust muscles with lots of connective tissue, so you need to hammer this area like you're tenderizing meat.

After spending some time on your trap, work your way down the border of your shoulder blade or scapula, stopping at each rib. As you can see in the photos below, the target area is from your trap (left) down the area bordering your scapula (right). Spend at least 30 seconds on each rib and accumulate 15 to 20 slow arm swings before switching sides.

Target Area:
Trap to border of scapula

1) Position a ball in your trapezius—the area above your scapula, between your neck and shoulder.

2) Reach your arms toward the ceiling, then drive your heels into the ground and elevate your hips to create additional pressure.

3) Keeping your hips elevated, move your arm over your head. Remember, you want to keep your shoulder in a stable position, so don't bend your elbow or internally rotate your shoulder as you go overhead.

4) Keeping your arm straight, reach across your body and try to touch your opposite hip.

5) Continuing to move your arm through its full range of motion, maneuver your hand underneath your back. From here, you can drop your hips to the floor, focusing on driving the ball into your trap and the border of your scapula.

Trap and First Rib Smash

When you sit, stand, or move in a position of upper back flexion, your shoulders and the surrounding tissues become adaptively stiff. More specifically, your first ribs, traps, and neck can become tight and restricted. Stiffness and poor function here can lead to neck, upper back, and shoulder pain. The Trap and First Rib Smash is one of the best mobilizations for restoring motion to your shoulder, neck, and trap complex. You can perform this mobilization in a doorway, in a corner of a wall, or on a post or beam. For the best results, use the contract and relax method and direct your breath into the area. Although it's not shown in the photos, you want to let your arm dangle as you press your weight into the ball. When you find a hotspot, maintain pressure while moving your arm around in as many directions as possible: across your body, behind your back, etc. You can also move your head to the side to tie in your neck.

**2 minutes
on each side**

1) Nestle a ball between your clavicle and the base of your trap and neck.

2) Keeping the ball pinned in place, drive your body into the ball. From here, contract by engaging your trap and shoulders and then relax, pushing the ball deeper into your neck and shoulder. After a few contractions, raise your arm overhead and move your hand behind your back. You can also use your hand to gently pull your head away from the ball. Keeping pressure into the ball, you can lower your upper body and allow the ball to roll over the top of your trap.

PRESCRIPTION 3
Chest and Anterior Shoulder

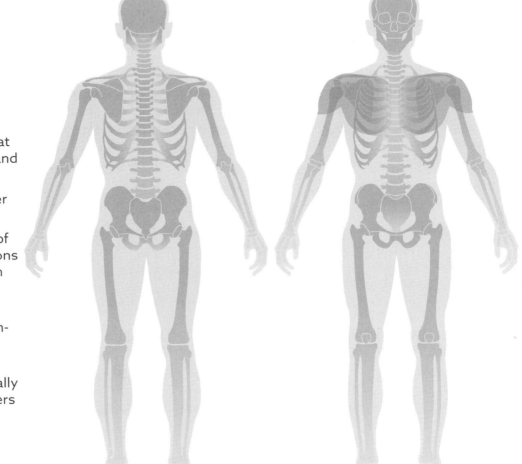

This prescription can be used to treat these symptoms and restrictions:

- Anterior shoulder pain
- Shoulder range of motion restrictions (internal rotation and extension)
- Stiffness from forward-head-on-neck position
- Stiffness from rounded (internally rotated) shoulders
- Tight chest

Methods:
- Contract and relax
- Pressure wave
- Smash and floss
- Tack and twist

Tools:
- Small ball
- Large ball

Total time:
12 to 16 minutes

Overview

Maintaining a solid posture is one of the hardest aspects of being deskbound. Whether it's caused by fatigue, being static for too long, or losing focus, it's easy to slip into a poor spinal shape. Here's a common example: Standing at your computer, you drop your head to look down at your keyboard and then forget to reset your position. Now you're standing with your head positioned out in front of your body. With the weight of your head pulling your upper body toward the floor, your shoulders collapse forward. Before you know it, you've been standing this way for 20 minutes. This prescription attacks the tissues around your chest and shoulders that become adaptively stiff as a result.

When you round your shoulders forward, three things happen. First, your spine rounds, creating a hinge at the base of your neck and upper back. Second, that rounded upper back makes it difficult for you to organize and stabilize your shoulders. Third, your tissues adapt to the position. So, if your shoulders are rounded forward, the tissues around the fronts of your shoulders, chest, and neck form a cast around that position, ultimately making it difficult to pull your shoulders back into a good position. This prescription helps restore function to those stiff tissues.

Use this prescription to attack the anterior, or front, of your shoulder and the next prescription to mobilize the posterior, or back, of your shoulder. Remember, you are a system of systems. To get the most out of your mobility work, always think about what is connected to the tissue or area that you're trying to change.

Anterior Trunk Mobilization

This technique is very similar to the Anterior Neck Mobilization from Prescription 1 (page 258), in that you use the power of your hands to pressure a ball into your body, then apply the tack and twist method to restore sliding surface function to the tissues underneath the ball. But instead of targeting your neck, the Anterior Trunk Mobilization targets your chest just underneath your clavicle, or collarbone. This area gets tacked down and junky in just about everybody. This is a great way to clean up the area and prepare yourself for the other mobilizations in this sequence.

**2 minutes
on each side**

Target Area:
Under clavicle, insertion near front of shoulder, and entire pectoral region

1) Your chest muscles attach to your clavicle. This is one of the target areas. To begin, press your fingers under your collarbone. You should feel the muscles inserting into that region. This is where you want to nestle the ball.

2) Place the ball underneath your clavicle, overlap your hands, and then press the ball into your chest.

3) Maintaining pressure, bind up the tissue by twisting the ball in place.

4) Tilt your head away from the ball, pull your shoulders back, and move your head in different directions.

Blue Angel

**2 minutes
on each side**

When it comes to restoring suppleness and undoing stiffness around the fronts of your shoulders and chest, the Blue Angel is king. What's great about this mobilization is that you can implement all the mobility methods—tack and twist, contract and relax, pressure wave, and smash and floss—which enables you to make a ton of change in a very short time. The good news is that you can do it facedown and respectfully hide your pain face.

To get the most out of this mobilization, you want to mimic the motion of making a snow angel with your arms. If lying on the ground is not an option or is too painful, you can perform this mobilization against a wall—see the Blue Angel Wall Option, opposite.

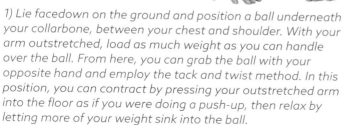

1) Lie facedown on the ground and position a ball underneath your collarbone, between your chest and shoulder. With your arm outstretched, load as much weight as you can handle over the ball. From here, you can grab the ball with your opposite hand and employ the tack and twist method. In this position, you can contract by pressing your outstretched arm into the floor as if you were doing a push-up, then relax by letting more of your weight sink into the ball.

2) Keeping your arm outstretched, externally rotate your shoulder while sliding your arm across the floor toward your head.

3) Move your arm toward your legs.

4) As you do this, internally rotate your arm so that your thumb is pointing toward your back.

5) To capture all the corners of your shoulder, place the back of your hand on your lower back.

6) With your arm behind your back, try to get more pressure over the ball by rotating your upper body away from the shoulder you're mobilizing.

Blue Angel Wall Option

You can also perform the Blue Angel against a wall or in a doorway. A doorway is ideal because you're free to move your arm out in front of your body, allowing you to hit nasty corners that are unavailable to you when you're mobilizing on the ground. The only caveat is that you can't get as much pressure or weight into the ball. Otherwise, use the same movements and methods as described above.

Lateral Opener

Once you've fed some love to the tissues surrounding the fronts of your shoulders and chest, they're primed for a dynamic mobilization, which focuses on lengthening and restoring normal range of motion to the shoulder joints and surrounding musculature. This technique restores normal range to those tissues that have become adaptively stiff and makes it easier to get into a neutral shoulder position. It's quick and easy and can be done just about anywhere.

**2 minutes
on each side**

1) Grab the edge of your desk or a door frame, positioning your hand slightly below shoulder height. If that is not an option, you can place your palm flush against a wall with your thumb facing upward. This ties in your anterior shoulder, biceps, and chest.

2) Keeping your belly tight and spine neutral, rotate your body away from the arm you're mobilizing.

3) Use your opposite hand to pull your head away from your extended shoulder. This ties in the components of your neck and shoulder complex.

Shoulder Extension Mobilization

**2 minutes
on each side**

This mobilization is similar to the Lateral Opener, but it creates a big stretch around the front of your shoulder rather than in your chest. Like the Lateral Opener, you can do it just about anywhere.

The goal with the Shoulder Extension Mobilization is to get your arm behind your back. You're working on getting your shoulder into extension (arm-behind-your-back range of motion), which counters rounded shoulders. As with any mobilization, you have to use a combination of motions and positions to bring about the most change.

1) Position your arm behind your back with your palm down. Your hand should be slightly below shoulder height and your shoulder in a neutral position.

2) Keeping your hand pinned in place, turn away from your arm and lower your elevation. For best results, contract and relax and try to turn and sink deeper into the stretch during the relax phase.

If you want to mobilize both arms simultaneously, perform the Sink Mobilization Option below.

Double-Arm Shoulder Extension (Sink Mobilization Option)

This is a great two-for-one option. It's just like the Shoulder Extension Mobilization, but you mobilize both arms simultaneously. You can do it on a fence pole or railing, gripping the inside of a sink, or with your hands on a desk, as shown—anything to which you can anchor your hands is fair game. Getting both hands behind your back can be difficult; if the position proves too challenging, stick with the single-arm option.

2 minutes

As in the single-arm option, position your arms behind your back, keeping your shoulders and posture neutral, and then lower your elevation to increase the tension.

PRESCRIPTION 4
Posterior Shoulder and Lat

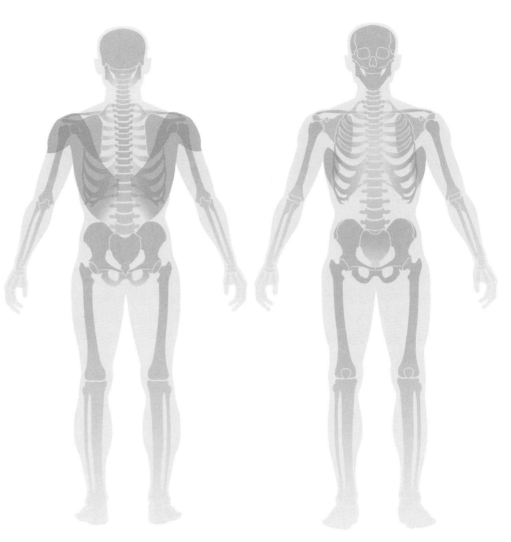

This prescription can be used to treat these symptoms and restrictions:

- Overhead range of motion
- Posterior shoulder pain
- Rotator cuff pain
- Shoulder external rotation range of motion
- Tight lats

Methods:
- Contract and relax
- Pressure wave
- Smash and floss
- Tack and twist

Tools:
- Small or large ball
- Roller

Total time:
10 minutes

Overview

Spending time hunched over the steering wheel of your car or your laptop will facilitate a set of predictable changes in your tissues. The external rotators of your shoulders, for example, will become what is known as "locked long." The muscles that internally rotate your arms and shoulders will become "locked short." It is far more effective for your body to stiffen down a habitually inefficient position than it is to try to actively support that position. The mobilizations in this prescription are intended to remedy the soft tissue changes in and around the backs of your shoulders.

Shoulder Rotator Smash and Floss

Habitually rounded shoulders can cause the external rotators of your shoulders to get overstretched, brittle, and stiff, which can lead to acute posterior shoulder pain. The good news is that restoring sliding surfaces and relieving pain is simple. All you need is a hard ball, like a lacrosse ball or softball, and you can effectively unglue those compromised tissues that may be causing you discomfort.

There are two ways to perform this mobilization: you can remain on your back (Option 1) or roll up onto your side (Option 2). Start with Option 1 and see how it feels. If you can handle more pressure, transition to Option 2.

**2 minutes
on each side**

Shoulder Rotator Smash and Floss (Option 1)

1) Position a small or large ball above the insertion of your lat near the back of your armpit. This is where the external rotators insert behind your shoulder.

2) With the tissues behind your shoulder tacked down on the ball, internally rotate your shoulder by dropping your hand to the floor.

3) Then externally rotate your shoulder by moving your arm toward your head. Slowly moving your hand back and forth like this 20 to 30 times, or for 2 minutes.

Shoulder Rotator Smash and Floss (Option 2)

1) From your back, position a small or large ball above the insertion of your lat near your armpit, then roll up onto your side.

2) Keeping your elbow bent, use your opposite hand to push your other hand toward the ground in the direction of your head. Don't be in a rush to move on. Contract and relax in this position and really try to get some weight over the ball.

3) Then internally rotate your shoulder by pushing your hand toward the ground in the opposite direction.

Overhead Stretch

This is a simple yet effective way to mobilize the tissues that are limiting your overhead position. The best part is, you can do it right at your desk. Remember, you always want to mobilize in a position that is similar in shape to the movement and position you're trying to change. So, if you're trying to improve your overhead position, it makes sense to put your arm(s) overhead and mobilize anything that might be limiting your range. The Overhead Stretch and Overhead Tissue Smash (pages 276 and 277) are perfect examples.

2 minutes

1) Place your hands palms down on your desk.

2) Stabilize your shoulders by screwing your hands into the desk and locking out your elbows. More specifically, screw your left hand into the desk in a counterclockwise direction and screw your right hand into the desk in a clockwise direction. Remember, you're not turning them outward; you're just exerting an outward force.

3) Keeping your belly tight and your arms locked out, walk your feet back while hinging forward from your hips. Your knees should be fully extended and your back flat. From here, you can contract and relax and try to push your chest to the ground. The key is to stay active by employing the contract and relax method.

Overhead Tissue Smash

This mobilization is a companion piece to the Shoulder Rotator Smash and Floss (pages 273 and 274). If your lats are tight, they will compromise your ability to raise your arms overhead and effectively stabilize your shoulders. Remember, the latissimus is a long muscle that runs from your armpit to your lower back.

As with most mobilizations that are performed on the ground, you can also perform this one against a wall.

1) Position a small ball in your armpit near the insertion of your lat and rotator cuff, on the back side of your shoulder.

2) Keeping as much weight as you can handle over your arm, slowly roll toward your belly by rotating your upper body. The idea is to pressure wave back and forth across your lat and shoulder.

3-4) After oscillating back and forth around your armpit area, move the ball down your lat. You want to work the entire length of your lat, focusing on the areas that feel tight and ropy.

Roller Option

You can also use a roller to perform this technique.

PRESCRIPTION 5
Low Back and Trunk

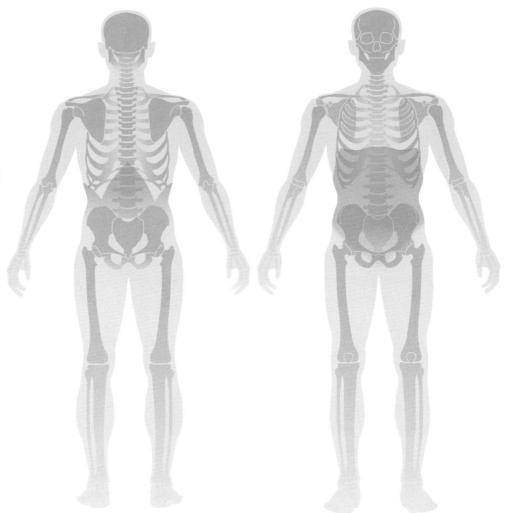

This prescription can be used to treat these symptoms and restrictions:

- Abdominal pain
- Compromised breathing
- Hip extension range of motion
- Hip pain
- Low back pain
- Sciatica
- Spinal rotation, flexion and extension range of motion

Methods:
- Contract and relax
- Pressure wave
- Smash and floss

Tools:
- Small ball
- Large ball
- Small bouncy ball or under-inflated soccer/volleyball

Total time:
14 minutes

Overview

This prescription is great for treating low back pain and supporting the hard-working muscles of your trunk.

We've established that poor spinal mechanics and sitting can cause adaptive stiffness and irritation in the discs, ligaments, and muscles around your spine and trunk. And when that happens, low back pain is often the result. Although there are other contributing factors to consider, like previous injuries, arthritis, obesity, and stress, we would argue that one of the leading causes of low back pain and trunk-related problems stems from poor posture, prolonged sitting, and a lack of basic self-maintenance. Having spent the majority of this book outlining a protocol for preventing and resolving the issue from a mechanical standpoint, let's turn our attention to the maintenance side of things.

This prescription targets the muscles that are responsible for keeping your spine braced, as well as the muscles that may get stiff when you move poorly or sit for too long.

In a perfect world, your spinal position would always be stellar, and you would never have to sit in a chair for an extended period. The reality is that you're going to have to sit sometimes, and there will be times when you become absent-minded and move, stand, or sit poorly. We're human, after all. This is why maintenance is a daily practice. It's a way of buffering poor mechanics and preventing dysfunction.

So, whether you're looking to treat low back pain or compromised breathing mechanics or just to perform some basic maintenance around your trunk, this is the prescription to employ.

Low Back Smash

This is one of the first mobilizations that we recommend to people who are suffering from low back pain and discomfort. It's quick and easy and offers immediate results. By sticking a small ball in your low back and upper gluteal region, you can effectively unglue the matted-down tissues that contribute to low back pain and restrict movement and positional quality. The goal is to work back and forth from the side of your hip to your spine, trying to stay on the crest of your pelvis and glutes.

You can perform this mobilization with your feet on the ground, as demonstrated on the following page, or with your feet propped up on a box or other elevated surface. Working on the ground is easier and less aggressive, but if you don't feel like you're getting enough pressure, experiment with elevating your feet. While it is more intense, we like the latter option because it's easier to mobilize with your spine neutral.

**2 minutes
on each side**

Target Area:
Upper gluteal region (left) and just above your hipbone, or iliac crest (right)

1) Place a small ball on your lower back, just above your pelvis. Focus on keeping your midline engaged to avoid breaking into overextension.

2) If you're mobilizing your left side, cross your left foot over your right leg. This will take up the soft tissue slack in your low back. If you're mobilizing your right side, cross your right foot over your left leg.

3) Slowly shift your hips toward your left. The goal is to roll back and forth, pressure waving across the tissues of your lower back and upper glute.

QL Side Smash

**2 minutes
on each side**

Like the erector spinae, the quadratus lumborum, or QL, can be a direct source of low back pain when prolonged sitting and poor spinal mechanics are the norm. Your QL muscles are essentially the hamstrings of your spine, with direct attachments onto each lumbar vertebra. When you're not in a neutral spinal position with your abs engaged (braced), the muscles that connect your spine to your pelvis and legs, like the QL, psoas, iliacus, and rectus femoris, pick up the slack. In other words, when you're not neutrally braced, these muscles have to work extra hard to keep you upright. Once stiffness in these tissue systems sets in, low back pain can easily be the result.

The goal with this mobilization is to sink into that QL, which runs from the bottom of your rib cage to the top of your pelvis, using a large ball such as a Supernova or softball. And don't feel limited to just your QL. Hunt for your stiff areas. They might include your erector spinae, which are the big muscles running down the sides of your spine, or your obliques, which are on the sides of your abdomen.

As with most mobilizations performed from the ground, you can also do this one against a wall while standing. The ground is a better choice, however, as the tissues supporting your spine won't be working against gravity.

1) Position a ball on the side of your lower back between your rib cage and hipbone.

2) Keeping your shoulders flush with the ground, elevate your hips and shift your body over the ball. From here, you can contract and relax, breathing into the area by focusing your inhalations into the ball and relaxing as you exhale. You can also pressure wave by making small oscillations over the QL and low back area and rotating your hips and dropping your leg(s) to the side.

3) Once you've hit your QL and erector spinae, roll onto your side and target your oblique. The idea here is to twist your torso over the ball and seesaw back and forth over your side, smashing your oblique, QL, and high glute. To make this particular position more effective, you can lengthen the tissues by extending your bottom arm overhead.

Gut Smash

2 minutes

The trunk is a hardworking system that is bound to get stiff, whether you're organized or disorganized. The simple fact is, you spend a lot of time and energy creating and maintaining abdominal stability, which causes the layers of your abdominal structure, particularly your psoas, to get stiff and ropy. And when tension builds through that system, it can compromise your breathing mechanics and trigger pain in your lower back. Think about the last time you had sore legs. What did you do for them—get a massage, roll them out, "stretch"? Now think about the muscular system that makes up the container of your trunk. Have you ever treated these tissues like the other muscles or soft tissue systems of your body? We bet not. Here we offer the first of two options for attacking the deep tissues of your trunk and abdominal musculature.

Global Gut Smash (Option 1)

As the name implies, the Global Gut Smash hits all the abdominal muscles and tissues surrounding your spine and trunk. This mobilization is especially important if you've had any kind of abdominal surgery, like a C-section or an appendectomy, which usually leaves behind layers of scar tissue.

To execute this mobilization, you will need a large ball with some pliability, like a kid's bouncy ball or an under-inflated volleyball or soccer ball. Remember, tissues are considered "normal" when you can apply moderate pressure without experiencing pain or discomfort. If you have visceral symptoms or achiness, it's an indication that your abdominal region is stiff and not working at full capacity. We recommend the Global Gut Smash as a first step; then use the Targeted Gut Smash to attack more specific hotspots and trigger points around your diaphragm, psoas, and abdomen.

Here Kelly is using Jill Miller's Coregeous ball, which is basically a mini physio ball. For best results, you need to penetrate into the basement level of your tissues by sinking all of your weight into the ball. To do this, take a big breath, hold it for a few seconds, and then exhale. As you breathe out, relax your weight over the ball and let it sink deeper into your abdomen.

1) Here Kelly is demonstrating the same technique with a slightly deflated volleyball. To begin, lie over the ball, positioning it between your hipbone and rib cage, to the outside of your belly button.

2) Add movement to the mobilization by sliding your knee to your hips.

3) You can also move your leg from side to side or pressure wave by twisting your upper body over the ball.

Targeted Gut Smash (Option 2)

1) Positioning a large ball between your belly button and hipbone (or at the border of your rib cage if you're targeting your diaphragm), lean over a stool, countertop, or box.

2) Folding your upper body over the ball, use your breath in conjunction with your weight to allow the ball to sink deeper into your belly. Once you've relaxed over the ball, floss by bringing your knee up.

3) You can extend your leg backward and to the side or twist your upper body over the ball for a different flossing effect.

PRESCRIPTION 6
Elbow

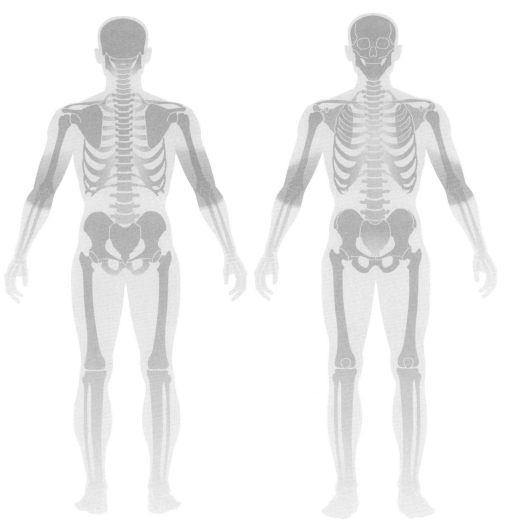

This prescription can be used to treat these symptoms and restrictions:

- Carpal tunnel syndrome
- Elbow pain
- Elbow range of motion (difficulty bending and straightening)
- Golfer's elbow
- Tennis elbow
- Triceps stiffness

Methods:
- Contract and relax
- Pressure wave
- Smash and floss
- Tack and twist

Tools:
- Small ball
- Roller

Total time:
12 minutes

Overview

When was the last time you dedicated 10 minutes to treating the tissues surrounding your elbows? Wait, we already know the answer: never.

People tend to put all their mobility focus into their hips and back, neglecting other areas of the body. And it makes sense. The hips have an important job, and people tend to spend a ton of time working on the primary movers connected to the hips, like the glutes, quads, and hamstrings. But what about the primary movers of the shoulders and arms? When it comes to mobilizing, most people spend very little time on their triceps and elbows. This is a mistake. Your triceps affect your shoulders like your quads affect your hips. If these regions get stiff, it's going to negatively impact your mechanics and eventually express itself in the form of dysfunction. The mobilizations in this prescription will help you address these crucial tissue systems.

Epicondylitis (tennis elbow) is a real problem among deskbound warriors. Let's say you work an eight-hour day in front of a computer. Even if you take frequent breaks, your elbows are stuck in a bent shape for most of that time. What happens? Well, your elbows gets tight. And when these tissues become adaptively stiff in that bent position, elbow pain is often the result. Imagine driving your car only in third gear. You could do it, but the load on your car would not be good. The same is true about using only a small range of motion in your elbows for long periods.

Look, your elbows are workhorses. Bending and straightening your arms is a big part of daily life. This prescription will help you address elbow pain associated with desk work (and manage the general wear-and-tear of daily life) as well as improve your shoulder mechanics. You need to talk to your elbows and triceps at least every other week. You can do these mobilizations right at your desk, so there are no excuses.

Triceps Smash

The triceps are easy tissues to access. All you need is a ball or roller and something on which to prop your arm. You can do the Triceps Smash at your desk during a break or anytime you feel tension building around your elbow or shoulder. If your elbows flare outward when you try to stabilize your shoulders, it could be that your triceps are tight and limiting your range of motion. In other words, elbows flaring outward during pressing and pulling motions is one of the compensatory patterns linked to tight triceps.

**2 minutes
on each side**

Although the smashing techniques that we demonstrate employ different tools, the techniques and methods remain the same: pressure wave from side to side across the tissue, contract and relax on tight spots, and floss by curling and straightening your arm.

When smashing your triceps, consider the area that you're trying to change. If your elbow is in pain or you want to improve elbow range of motion, focus your attention around your elbow. If your shoulder hurts or you're having trouble stabilizing your shoulder in a neutral position, get on the ground and work on the long head of your triceps with your arm straight.

When you're rolling out your triceps on the ground, you'll probably feel a definitive clunk as you roll across the tissue. Although the clunk is common, it's not normal. You don't have an extra triceps bone.

Triceps Smash: Ball Option

1) Position the insertion of your triceps (right above your elbow joint) on a ball with your arm slightly extended. To take up the skin slack at your elbow, twist the ball toward you. Place your opposite hand on your biceps and apply downward pressure.

2) Employ the smash and floss method by curling your arm.

Triceps Smash: Roller Option

1) To use a roller, lie on your side, positioning the roller underneath your triceps near your shoulder.

3) Maintaining downward pressure, extend your arm. The idea is to bend and straighten your arm over the target area. You can also employ the contract and relax method in this position.

4) To get full movement of the tissue, bend your arm and then drop your palm toward the desk.

5) Continue to pressure wave across your elbow by rotating your arm in the opposite direction.

2) Keeping your arm extended, pressure wave across your triceps by rotating your arm and upper body toward the floor. The idea is to work back and forth across the tissue in this manner until you experience change or stop experiencing change. The contract and relax and smash and floss methods are also effective here.

Lateral Elbow Tack and Twist (Tennis Elbow)

**2 minutes
on each side**

This is a simple and quick mobilization for treating acute elbow pain. If you're deskbound, you can assume that your elbows, forearms, and wrists are working overtime with little compensation. When stiffness accumulates, elbow pain is often the result. "Tennis elbow" (formally called lateral epicondylitis) is a common term for this condition because it is typical among tennis players. But you don't have to play tennis to get tennis elbow. In fact, most of the people we see who are afflicted with this problem have never played tennis. It should be called "desk jockey elbow."

If your elbow aches or you are suffering from epicondylitis, this mobilization should be one of your first stops. Notice in the photos below that Kelly is targeting the top of his forearm near his elbow. When this area gets stiff and sore, it tends to cause a lot of discomfort. Like most of the techniques that we demonstrate, you can do this mobilization right at your desk using a small ball such as a lacrosse ball. If you type all day with your arms stuck in a flexed position, bookmark this mobilization and commit to doing it often.

Elbow Tack and Twist (Option 1)

1) Press a small ball into the top of your forearm next to the crook of your elbow.

2) To take up all the skin and soft tissue slack, twist the ball while maintaining downward pressure.

3) Flex your wrist and add rotation by internally and externally rotating your wrist and elbow.

4) Extend your wrist and splay your fingers. The idea is to create as much movement as you can by moving your wrist and hand in all directions.

Elbow Tack and Twist (Option 2)

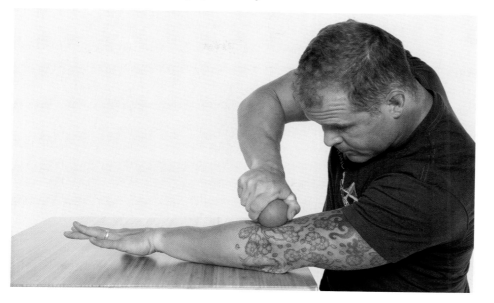

To get more downward pressure into your elbow and forearm, try propping your arm up on your desk. This removes the counterpressure that you need to apply when performing Option 1 and allows you to drive the ball deeper into your arm.

Medial Elbow Tack and Twist (Golfer's Elbow)

The inside of the elbow is another area that can quickly become an epicenter of dysfunction. Doctors sometimes refer to pain in this area as "golfer's elbow," which is basically the same thing as tennis elbow, but instead of causing pain on the outside of the arm, it causes pain along the inside of the arm.

As with tennis elbow, golfer's elbow is not limited to golfers. This mobilization can provide relief and prevent pain and restrictions around the inside of your elbow.

**2 minutes
on each side**

1) Press a small ball into the inside of your elbow, right above the bony bump.

2) Applying pressure, take up all the skin and soft tissue slack by twisting the ball into your arm.

3) Floss by bending and straightening your arm. For best results, continue to tack and twist and smash and floss around the inside of your elbow.

PRESCRIPTION 7
Forearm, Wrist, and Hand

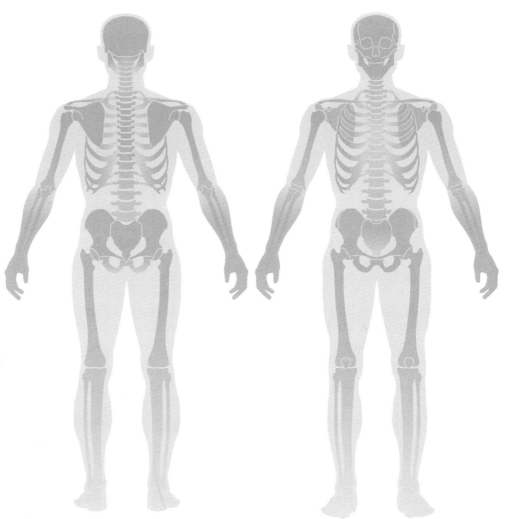

This prescription can be used to treat these symptoms and restrictions:

- Carpal tunnel syndrome
- Elbow pain
- Thumb and hand pain ("text thumb")
- Wrist pain
- Wrist range of motion (flexion and extension)

Methods:
- Contract and relax
- Pressure wave
- Smash and floss
- Tack and twist

Tools:
- Small ball
- Peanut tool
- Large ball
- Marker or highlighter

Total time:
12 minutes

Overview

Wrist pain and carpal tunnel syndrome are all-too-common problems in deskbound society. The greatest misunderstanding about wrist and hand pain is that many factors contribute to the development of this devastating condition. Up front, we tend to focus on the carpal tunnel, that bony tunnel in the wrist, as the chief culprit. While poor mechanics can certainly compromise the nerves, tendons, and vasculature running through this bony canal, the beautiful structures of your wrist didn't cave in overnight. You have to think about the fact that your nervous tissue originates upstream and runs through a lot of meaty tunnels on the way to your wrists and hands. Can you remember whether your typing skills class included a body posture component? Did your typing teacher teach you about maintaining stable shoulders so that your elbows and forearms don't have to bend or twist to compensate? How about forearm stiffness and basic range of motion drills? Our guess is that you received none of this training. Instead, you flung your seated body in front of the keyboard and started cranking out the great American novel. And because no one ever explained that stiffness, weakness, numbness, and tingling are signs that you need to move, you were able to blow through all of your body's natural stop signs.

If we were to perform a forensic investigation into the causes of wrist and hand pain, we'd likely turn up at least 10 contributing factors, including the forward-head-on-neck position, a stiff thoracic spine, internally rotated shoulders, and stiff forearms. We haven't even talked about the sustained pressure on your wrists at the keyboard and the fact that hypo-hydration makes this area a sticky mess. A complete treatment plan has to include improving neck and head positioning as well as freeing up the systems through which the nerves from your neck to your hand run. Oh, and did we mention how that stress-induced neck-breathing pattern causes your neck flexors to make a nerve sandwich between your first rib and scalene muscle? It may seem like a lot to fix, but the road out of wrist hell is paved with improving your body's mechanics.

Texting, scrolling, and other smartphone activity can also lead to something that is being referred to as "text thumb," which is an overuse injury that causes thumb pain and stiffness.

You can prevent and eliminate these symptoms by correcting your position and performing daily basic maintenance. When it comes to typing, make sure to take frequent breaks and get some motion into your arms and wrists. If you need an ergonomic keyboard and/or mouse, go for it. To make holding your phone easier, consider attaching a PopSocket (www.MobilityWOD.com/product/popsockets/), which is a protruding grip that sticks to the back of your mobile device and makes it easier for you to grip it without compromising your wrist or thumb position. But don't use these tools as crutches. You still need to move and focus on maintaining a neutral hand and wrist position.

To keep your wrists free of dysfunction and avoid the symptoms associated with carpal tunnel syndrome and text thumb, you need to address the hardworking muscles of your forearms. Simply stretching is not going to do much for you. We like to point out to our patients that we've never met a piece of beef jerky that responded to stretching. If you suffer from carpal tunnel syndrome or wrist and thumb pain, use this prescription to treat the symptoms and get back to baseline.

Forearm Smash

**2 minutes
on each side**

If your elbow or wrist is in pain, you need to look upstream and downstream and attack the tissues that are pulling on your elbow and hand. Your forearms do an unbelievable amount of work for your hands, and you probably do nothing for them. If the strings that control your hands have turned into steel cables, don't be surprised when your elbows, wrists, and hands glow with pain.

Like most of the techniques that we demonstrate, you can do this one right at your desk. The idea is to cover the entire posterior compartment of your forearm, from elbow to wrist.

As you can see in the photos, you have a couple of options. You can position a ball or peanut tool under your forearm and then press into your arm using a small ball, as demonstrated in Option 1 below. This allows you to target the anterior and posterior compartments of your forearm simultaneously, making it a nice two-for-one mobilization. You can also employ the Forearm Extensor Smash (Option 2), which targets the top of your forearm.

Forearm Stack and Smash (Option 1)

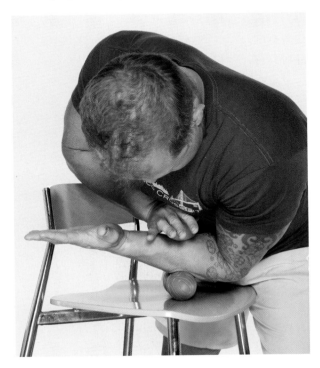

Position your forearm over a small ball or peanut tool (Kelly is using a Gemini here) with your palm facing the ceiling. Then position a small ball over the bottom tool. Drive the top ball into your forearm by shifting your weight over your arm and pressing down with your opposite hand. From here, you can contract and relax by splaying your fingers or making a fist, and smash and floss by flexing and extending your wrist.

Forearm Extensor Smash (Option 2)

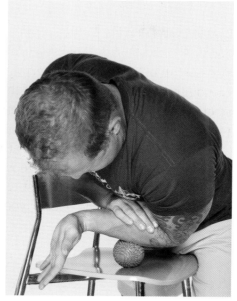

1) Position your forearm over a small or large ball with your palm facing the ceiling.

2) To get enough pressure into your forearm, shift the weight of your upper body over your arm while pressing your opposite hand into your forearm, directly over the ball.

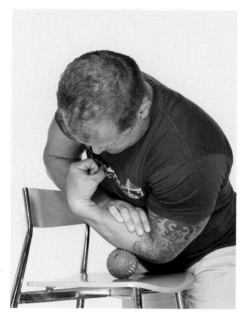

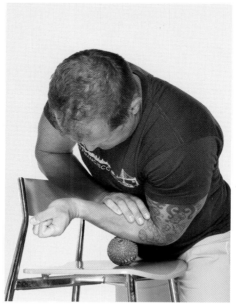

3) With the weight of your body over your arm, employ the smash and floss method by curling or flexing your wrist. You can do this with your hand open or closed.

4) Continue smashing and flossing by extending your wrist. You can also contract and relax by splaying your fingers or making a fist.

Wrist Sequence

**2 minutes
on each side**

In addition to causing repetitive stress injury, spending too much time on your phone or keyboard can cause your skin to get tacked down to the underlying tissue and bone, restricting movement and compressing nerves. The next time your thumb or wrist hurts, use this technique in conjunction with the other mobilizations in this prescription to restore your sliding surfaces. The key is to cover the entire circumference of your wrist.

1) Position a small ball on the outside of your wrist.

2) Applying pressure, take up the skin slack by twisting the ball.

3) Extend your wrist for a flossing effect.

4) Continue smashing and flossing by curling or flexing your wrist toward your body.

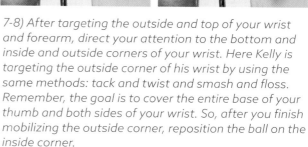

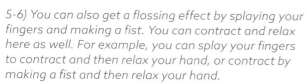

5-6) You can also get a flossing effect by splaying your fingers and making a fist. You can contract and relax here as well. For example, you can splay your fingers to contract and then relax your hand, or contract by making a fist and then relax your hand.

7-8) After targeting the outside and top of your wrist and forearm, direct your attention to the bottom and inside and outside corners of your wrist. Here Kelly is targeting the outside corner of his wrist by using the same methods: tack and twist and smash and floss. Remember, the goal is to cover the entire base of your thumb and both sides of your wrist. So, after you finish mobilizing the outside corner, reposition the ball on the inside corner.

Thumb and Hand Sequence

2 minutes on each side

Smartphones—like chairs—are not really designed with our human structures in mind. Consider your wrist position and the number of thumb cycles you burn through when using your phone: you use your thumbs to scroll, push, and type with your wrists locked in an untenable position. The problem is that mobilizing your thumb and hand is not exactly easy. You have to find clever ways to deal with your hotspots. For example, on the next page Kelly is using a highlighter to smash into his thumb pad. A lacrosse ball will work in a pinch, but a tool with a small, flat, hard surface is a much better option. If you have a highlighter or marker lying around, put it to use by smashing the hotspots around your thumb and hand. Otherwise, grab your trusty lacrosse ball and get some work done.

Wrist and Thumb Tack and Twist

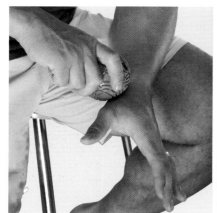

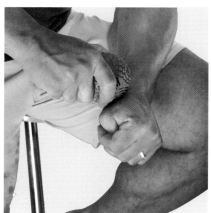

1) Position a small ball on the inside corner of your wrist.

2) Applying pressure, take up the skin slack by twisting the ball.

3) Make a fist over your thumb.

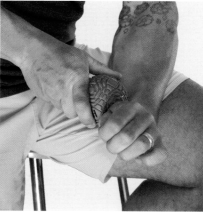

4-5) With your fingers clenched around your thumb, move your wrist around in all directions for a flossing effect.

Thumb and Hand Smash and Floss

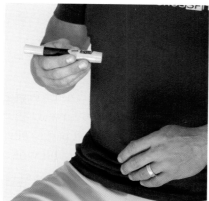

1) For this mobilization, find a marker or highlighter with a flat, hard bottom.

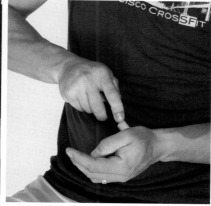

2) Drive the bottom of the highlighter into your palm, near the base of your thumb.

3) Extend your wrist and splay your fingers and thumb for a flossing effect.

4) Still driving the highlighter into the base of your thumb, wrap your palm around the tool. Notice how this subtle movement changes the angle and pressure being applied to your thumb.

5) From here, you can splay your fingers for a slightly different flossing stimulus.

6) After mobilizing around the base of your thumb, target the opposite side by driving the highlighter into the top of your hand between your thumb and index finger.

7-8) Contract and relax and smash and floss by opening and closing your hand.

PRESCRIPTION 8
Glutes

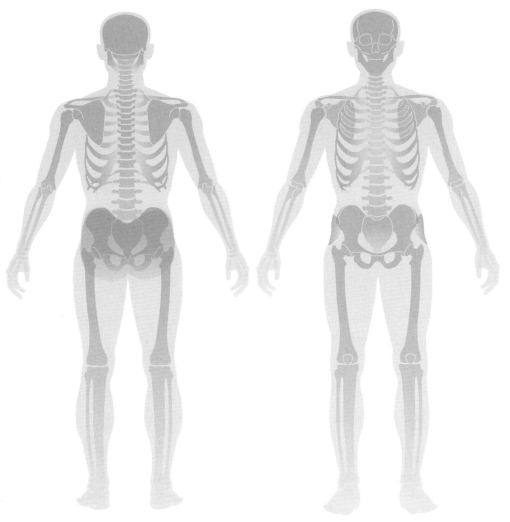

This prescription can be used to treat these symptoms and restrictions:

- Hip flexion range of motion
- Hip impingements
- Sciatica
- Squatting range of motion
- Tight and stiff pelvic musculature (chair butt)

Methods:
- Banded flossing
- Contract and relax
- Pressure wave
- Smash and floss

Tools:
- Small ball
- Large ball
- Band

Total time:
12 minutes

Overview

This prescription is designed to unglue the laminated tissue in your posterior. If you get trapped in a chair for an extended period, this prescription will help you deal with the symptoms. It is also a great prescription to employ when you know that you're going to be stuck in a seated position—say, on a long road trip or flight. Perform this prescription both before and after your time in the chair.

Glute Smash and Floss

This is the best mobilization for restoring sliding surface function to your glutes. You'll be shocked at what a difference it makes. To test and retest, simply mobilize one cheek for two minutes, then stand up and squeeze your glutes as hard as you can. You'll find that the side you mobilized contracts with a lot more force than the other, which is a sign that you've just increased the efficiency of this muscle. In other words, you just made your glutes stronger and more powerful by reducing the internal resistance within that contractile tissue.

As with other large muscle group mobilizations—like the Quad Smash on pages 313 and 314—you don't need to worry about covering the entire glute region in one session. Just find a spot that is tight and painful (it probably won't take long) and then use the three sliding surface methods—pressure wave, smash and floss, and contract and relax.

As you can see, you can perform this mobilization in a chair (Option 1) or on the floor (Option 2).

**2 minutes
on each side**

Chair Glute Smash (Option 1)

The great thing about this mobilization is that you can do it right in your chair. Simply sit on a small ball, find a tight spot, and then smash and floss and contract and relax. To increase the intensity, cross your foot over your opposite leg. So, if you're mobilizing your left butt cheek, place your left ankle on your right knee.

Floor Glute Smash (Option 2)

1) Position a small ball on one side of your glutes.

2) Using one or both hands to help support the weight of your upper body, shift your weight over the ball and hunt for a tight spot.

3) With the ball tacking down the underlying muscle, move the surrounding tissue around by externally rotating your leg and dropping your knee to the floor. In addition to smashing and flossing with internal and external rotation of your leg, slowly roll from side to side. If you stumble across a particularly painful area, contract and relax until you get to the bottom of the tissue.

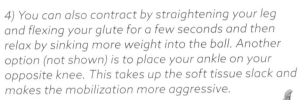

4) You can also contract by straightening your leg and flexing your glute for a few seconds and then relax by sinking more weight into the ball. Another option (not shown) is to place your ankle on your opposite knee. This takes up the soft tissue slack and makes the mobilization more aggressive.

Side Hip Smash

This is a great companion mobilization to the Glute Smash and Floss. As the name implies, it targets the side of your hip. Like the rest of the techniques in this prescription, the Side Hip Smash improves the mechanical efficiency of your hips. Although Kelly is using a large ball in the photos, you can use a small ball for a more targeted effect.

**2 minutes
on each side**

1) Position a large or small ball on the outside of your high glute, just below your hipbone.

2) From here, you have a couple of options. You can floss by bringing your knee up toward your belly, as shown; contract and relax by flexing your glutes and straightening your leg; or pressure wave by rolling toward your belly, smashing the tissue across the muscle fiber.

Hip Capsule Mobilization

**2 minutes
on each side**

As you may recall from Section 5, sitting at the back of a chair with your weight distributed over your hamstrings displaces the heads of your femurs into the tops and fronts of your hip sockets. Couple this change in the mechanical orientation of your hip joints with chronic, sustained seated hip positioning, and your hip capsules will very likely become very tight and stiff.

This is a great mobilization for resetting your hips to their biomechanically optimal orientation, clearing hip impingements, and reclaiming hip flexion range of motion. By aligning your knee directly underneath your hip and loading your weight over your femur, you can displace the head of your femur back into the posterior capsule of the joint. It's a quick and easy way to improve hip function without having to see a physical therapist.

1) Kneel on the ground and extend one leg behind you. Then shift the majority of your weight onto your grounded knee. The key is to align your knee directly underneath your hip and load your weight over your femur.

2) Keeping your weight over your knee, sit your hips back in the direction of your grounded leg. Imagine trying to pop the head of your femur out the side of your butt. Direct your line of force through your femur.

3) To increase the intensity, swing your grounded leg across your body and pin it in place using your opposite knee. From here, sit your weight back as demonstrated in the previous photo.

4) You can also work on internal rotation by kicking your grounded leg out to the side and sitting your weight back.

5) If possible, wrap a band around your hip crease and create a lateral or posterior distraction or tension. This will clear any impingement you may have in the front of the capsule. Note that if we spent more of our modern life working from our knees, this position would occur naturally.

PRESCRIPTION 9
Hip

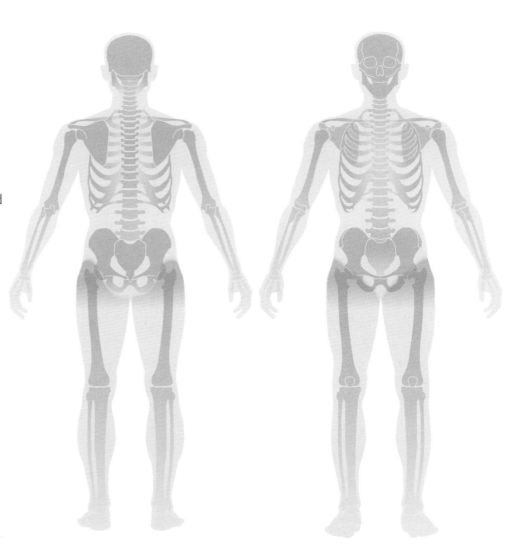

This prescription can be used to treat these symptoms and restrictions:

- Hip impingements
- Hip pain
- Hip range of motion (flexion, extension, and external rotation)
- Low back pain
- Squatting range of motion

Methods:
- Banded flossing
- Contract and relax
- Pressure wave
- Smash and floss

Tools:
- Large ball
- Band

Total time:
12 to 16 minutes

Overview

When it comes to sitting-related mobility issues, having tight anterior hips is the most common. As the Skin-Pinch Test on page 222 demonstrates, you lose the ability to stabilize your pelvis in a neutral position when your hips adapt to the sitting shape. As a result, your pelvis dumps forward and you sit, stand, walk, and move with an overextended back. Back pain and compromised movement patterns are to be expected. This prescription aims to correct the problem.

Whether you're trapped at an airport, watching a movie at home, or working at your desk, this prescription is a highly effective way to reclaim and maintain range of motion and reduce muscular stiffness in the fronts of your hips, as well as help resolve low back pain. Perform the techniques in this prescription anytime your back hurts or your hips feel tight. If you're trapped in a chair, this is one of the best prescriptions to implement throughout the day or at the end of your chair session.

Anterior Hip Smash

The Anterior Hip Smash targets the hip flexor "wad," or wad of muscle near the front of your hip. This is one area of your hips that is horribly affected by sustained sitting postures. By lying over a ball and employing the contract and relax, pressure wave, and smash and floss methods, you can restore suppleness and range of motion to this area. This technique is especially great for combating pain around the front of the hips.

**2 minutes
on each side**

Target Area:
*Anterior hip region/
hip flexor area*

Position a large ball on your hip flexor. From here, you can contract and relax or flex and extend your lower leg. You can also smash and floss by curling your heel toward your butt and moving your leg from side to side. Or you can pressure wave across your hip flexor area by rotating your hips from side to side.

Classic Hip Extension

**2 minutes
on each side**

This is an oldie-but-goody mobilization for opening up your hips. It targets your hip flexors and primes your hips for the infamous Couch Stretch (opposite). The best part about this classic kneeling hip opener is that you can do it right at your desk.

1) Kneel on the ground, distributing your weight over your grounded knee. If you're mobilizing on a hard surface, consider placing a towel or soft pad under your knee.

2) Maintaining a neutral spinal position, squeeze your butt and shift your weight forward. A lot of people mistakenly overextend at the low back as they open their hips. To avoid this mistake, keep your butt squeezed to stabilize your lumbar spine, keep your abs on tension, and focus on shifting your weight forward over your grounded knee. From here, you can contract and relax your "stretched" thigh to sink deeper into the movement. To increase the intensity and tap into your high hip stiffness, bring your left arm over your head (if your left knee is on the ground), lean back (but do not arch), and then come back to center.

Banded Hip Extension Option

The problem with the Classic Hip Extension is that it does not account for joint capsule restrictions. To make this mobilization more effective, hook your leg through a band and create a forward distraction while kneeling. The same rules apply: neutral spine and butt squeezed. The force from the band will pull your femur to the front of the socket and help improve the position and function of the joint.

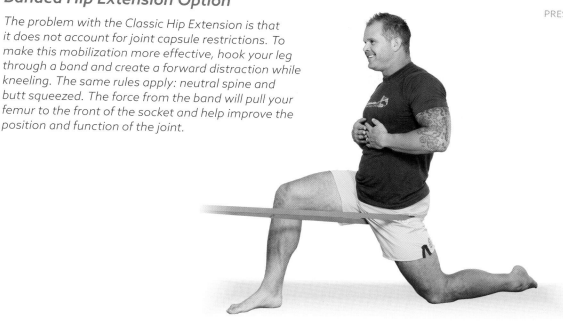

Couch Stretch

This is one of our favorite/least favorite mobility techniques in the book. It's our favorite because it opens up the hips while mobilizing the quads, which are two areas that become freakishly stiff from sitting. But it's our least favorite because it is brutal. Keeping your spine neutral and your butt squeezed—two critical components of performing the technique correctly—is like a mini workout.

We call this the Couch Stretch because Kelly created it while mobilizing against a couch. He was trying to figure out how to open his hips with his leg in full flexion. Performing this mobilization in front of the TV turned out to be quite beneficial because it took his mind off the pain and kept him from blacking out and vomiting on the living room floor. Okay, it's not that bad, but the distraction of TV does help.

If there's one mobilization that deskbound people should commit to doing daily, this is probably it. Most people are missing the critical ability to extend their hips efficiently. Exercise equipment like spin bikes and elliptical machines pander to this problem by never requiring the hips to fully extend. People will literally go for months without extending their hips beyond what is required for walking. So grab a bookmark and tag this page. We like to tell people that for every hour you spend sitting, you should spend two minutes mobilizing each hip in the Couch Stretch.

**2 minutes
on each side**

1) On your hands and knees, back your feet up to a wall.

2) Slide one leg back, driving your knee into the corner where the wall meets the floor and positioning your shin and foot flush against the wall.

3) Post up on your other leg, keeping the shin of your lead leg as vertical as possible. If you're unable to post up because you're too stiff, position a small box or chair in front of you for added stability.

4) Squeeze the glute of your rear leg and drive your hips toward the floor while maintaining a flat back.

5) After hanging out in the previous position for a minute or longer, raise your torso into the upright position. Again, if you find it difficult to support the weight of your body in the upright position, place a box or chair in front of you for added stability. Don't forget to contract and relax while performing this crucial mobilization. You have to involve your brain in the changes you are trying to make.

Deskbound Variations

Single Leg Flexion and External Rotation

This mobilization is great for improving squatting range of motion and mechanics. As you can see in the photos below, you're basically in a deep squat with one leg. Remember, when it comes to improving range of motion, the key is to mobilize in a shape that is similar to the position you want to change. This gives your mobility session purpose and intention. If you do this mobilization, you will squat better.

    The key to performing this technique correctly is to try to discover the vector or range where you feel restricted. Move your upper body in both directions, shoving your lead knee out, and making small circles with your hip. Explore and see what you can find.

**2 minutes
on each side**

1) Starting on your hands and knees, step one leg forward, keeping your lead shin vertical. Your back should be flat, your hips square, and your lead foot straight.

2) Keeping your lead foot stapled to the floor, allow your lead knee to drop to the side. As you do this, actively drive your hips toward the ground. From here, imagine drawing small circles on the floor with your hips.

3) To emphasize hip external rotation range of motion, twist your upper body away from your leg. If you have the mobility, try pushing your lead knee outward.

4) For a slightly different stimulus, rotate your upper body toward your lead leg. If you have the mobility, drop to your elbow and hug your knee. Although not shown here, you can also drive your hips back, straightening your lead leg, and then drop back into the position. This is a great way to tie in your hamstrings.

Banded Hip Option

This mobilization works well without a band. But if you can use a band, you should. It will help clear hip impingements and allow you to tap into deeper hip capsule restrictions.

Deskbound Variations

Sometimes mobilizing on the floor is difficult, especially if your hips are brutally tight or you're at the office. In either situation, mobilizing with your foot on a stool is an excellent alternative. With your lead leg elevated, you're not limited by your trailing leg. Aside from elevating your foot, the technique is the same: mimic the mechanics of the squat (foot straight, back neutral) and be active by moving your body into different positions.

Upper Leg

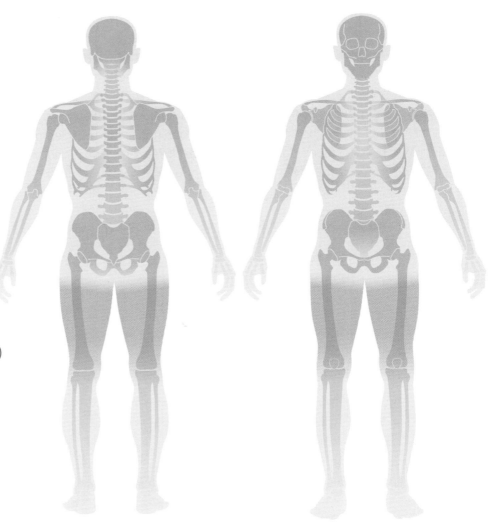

This prescription can be used to treat these symptoms and restrictions:

- Hamstring range of motion
- Hinging range of motion
- Hip pain
- Hip range of motion (flexion, extension, and external rotation)
- IT band syndrome
- Low back pain
- Sciatica
- Upper leg stiffness (quads, adductors, and hamstrings)

Methods:
- Banded flossing
- Contract and relax
- Pressure wave
- Smash and floss

Tools:
- Roller
- Small ball
- Band

Total time:
18 minutes

Overview

Just about everyone can benefit from mobilizing the large muscle groups of the upper leg. Your quads, hamstrings, and adductors are quiet workhorses in any situation and get adaptively stiff from prolonged sitting. In this prescription, we demonstrate a few techniques for restoring functionality to these hardworking muscles.

Quad Smash

The one problem with quad smashing is that the muscles of the upper leg are large and robust. You probably won't cover all the areas of your upper leg in one session. So you can attack it in a couple of different ways: you can go after a chunk, like the high region near your hip or the low region near your knee, or you can focus on the front, inside, or outside seam.

A common mistake is to quickly roll up and down the leg. We prefer to pancake and pressure wave laterally across the muscles. You'll probably need at least five minutes per leg. If you can't handle the full weight of your body over a roller, it's a good indication that your upper leg muscles need attention. When your knees don't hurt, your back feels amazing, and your legs feel fresh, you'll be glad you did it.

**5 minutes
on each side**

1) Lie on your side over a roller, positioning it directly underneath your leg. Here Kelly is starting near his hip on the outside, or lateral seam, of his leg and rolling toward the inside of his leg, but you can start anywhere. Notice that he is using his arms to support the weight of his upper body.

2) The IT band is a sensitive spot and cannot actually be stretched. You can, however, affect the muscle tissue that uses the IT band as an attachment. If you find a knot or tight spot, floss by curling your heel toward your butt. Contracting and relaxing over the problem spot is also vital.

3) Pressure wave across the length of the muscle by slowly rolling toward your belly.

4) Once you've found an area that needs attention, try flexing your leg back and forth over that area. You can also contract and relax by flexing and relaxing your quad.

5-6) Continue to pressure wave across your upper leg by slowly rolling toward your belly. From here, apply all the same smashing methods: pressure wave, contract and relax, and smash and floss. Stiff adductors (the muscles of the inner thigh) can also cause a host of problems. Your adductors are like the hamstrings of your groin.

Hamstring Ball Smash

When it comes to sitting upright with neutral posture, sitting at the edge of your seat is ideal, as demonstrated in Section 5. The problem is that it is not possible in all situations. Keeping your knees flexed while seated for 10-plus hours a day would be like keeping your elbows bent at 90 degrees. How would your arms feel when you tried to straighten them? Add the fact that these hard-working muscles are regularly being crushed under your entire body weight in the chair, and you have a recipe for grilled cheese tissue sandwiches.

As you can see in the photos, you can do this mobilization in a stool or a chair. Any hard, elevated surface will work. Sitting during this mobilization allows you to get more weight over the ball and gives you the ability to move your legs while mobilizing.

Like the Quad Smash, you will probably have to tackle this one in chunks. Focusing your attention on your lower hamstrings is great for combating knee pain. Cleaning up the area around your high hamstring, where it inserts into your pelvis, can provide a ton of relief from low back pain.

**2 minutes
on each side**

1) Position a small ball under your leg in the meat of your hamstring. You can place it near the insertion near your groin and glutes or in your lower hamstring next to your knee. The key is to find a hotspot. Once you find it, floss by curling your leg underneath the chair or stool.

2-3) You can also cross your leg over your knee and straighten your leg for different flossing effects.

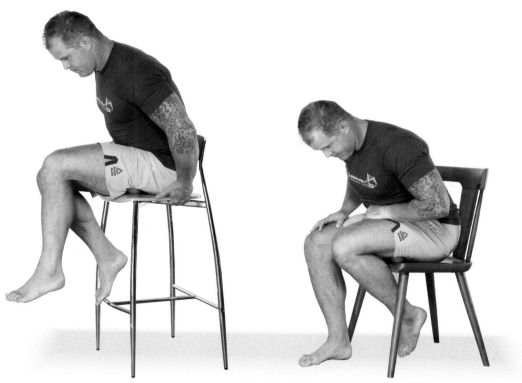

4) For more pressure, lean your upper body forward. And don't forget to contract and relax over tight and sticky spots.

Hamstring Mobilization

The Hamstring Mobilization improves hamstring range of motion and hinging mechanics. If you took the hip hinge test on page 226 and realized that you're missing capacity in your posterior chain, you should commit to performing this technique often.

    As you can see in the photos on the following two pages, you have a couple of options for mobilizing your hamstrings. The Banded Hamstring Floss (Option 1) is the best. Along with the Couch Stretch (pages 307 and 308), this is our go-to mobilization after long car rides and flights. The Classic Hamstring "Stretch" (Option 2) is great for mobilizing your hamstrings, but without the additional assistance of a band, you won't be able to tap into hip capsule restrictions.

**2 minutes
on each side**

Banded Hamstring Floss (Option 1)

1) Hook your leg through a band, wrapping it around your hip. Then create tension by walking forward and hinging forward from the waist. Notice that Kelly is in a sprinter's stance with his free leg slightly in front and his banded (rear) leg bent. If you can't reach the ground without rounding your back, position a chair or stool in front of you.

2) Keeping your back as flat as possible, create a flossing motion by repeatedly straightening and bending your banded leg and driving your hips back.

Classic Hamstring "Stretch" (Option 2)

1) Loop a band, strap, or belt around the bottom of your foot, then pull your knee to your chest. If you don't have a strap, hook your arm around the back of your knee. If you can reach your foot without sitting all the way up, grab the outside of your foot or hook your index and middle fingers around your big toe.

2) Keeping your back flush with the floor, pull your foot toward your head and straighten your leg. If you're not using a band, straighten your leg while maintaining a strong grip around the back of your knee. Your arm shouldn't move while you execute this option. From here, you can continue to floss by bending and straightening your leg, or contract and relax by resisting into your arms.

3) Scour around for tight corners by dropping your leg to the side. If you are holding your leg with your arm, wrap your fingers around your big toe or grab the outside of your foot. You can bend and straighten your leg in this position as well.

4) Keeping your hips flush with the ground, maneuver your leg toward the opposite side of your body. Again, from this position you can floss by bending and straightening your leg and contract and relax by resisting into your arm or the band.

PRESCRIPTION 11
Knee

This prescription can be used to treat these symptoms and restrictions:

- Knee pain
- Knee range of motion (flexion and extension)
- Upper and lower leg stiffness (lower quads and hamstrings, upper calves)

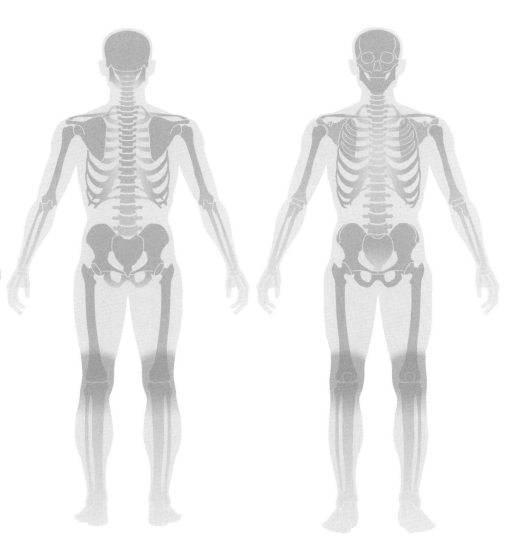

Methods:
- Contract and relax
- Pressure wave
- Smash and floss

Tools:
- Small ball
- Large ball
- Peanut tool

Total time:
10 minutes

Overview

If you have knee pain, you should be mobilizing the tissues surrounding your knee joints. That is exactly what this prescription shows you how to do. Consider this a catchall prescription for maintaining healthy knee joints, resolving knee pain and injuries, and improving knee flexion (bending) and extension (straightening) range of motion.

Suprapatellar Smash and Floss

The best way to improve knee mechanics and mobility is to feed some slack to the kneecap system by cleaning up your suprapatellar pouch, which is the area just above your kneecap (see "Target Area" below). Sitting in a chair forces you to keep your knees flexed, and this bent-knee position causes the tissues above your kneecaps to get stiff. This area of your knee joint is the reservoir for the slack required to fully bend your knee. Thus, when the area is tight, it creates a mechanical tensioner throughout the entire knee joint, including the kneecap. Mobilizing this area will improve your knee mechanics and unload your over-tensioned knee structures.

**2 minutes
on each side**

Target Area:
*The entire area above
your kneecap, from the
inside to the outside*

1) Lie belly-down on the floor and position a small ball just above your kneecap.

2) If you encounter a hotspot, you can contract and relax by flexing and relaxing your quad, or floss around the stiff tissue by curling your heel toward your butt.

3) Create a pressure wave by internally rotating your leg. From here, you can contract and relax or floss by flexing your knee.

4) Continue to smash across your suprapatellar pouch and quad tendon until you reach the lateral part of your knee. Employ the contract and relax and smash and floss methods again from this position.

Knee Scissor Smash

The Knee Scissor Smash is a simple yet effective technique for cleaning up the area along the inside of your knee (vastus medialis, or teardrop muscle). This technique can be done in a couple of different ways. If all you have is a small or large ball, stick it between your knees and scissor your legs back and forth. With the ball positioned between your legs, you can effectively smash both legs simultaneously, making it a nice two-for-one mobilization. If you have a ball and a peanut tool, you can turn this into a three-for-one mobilization by placing the peanut tool under your bottom leg.

2 minutes

Knee Scissor Smash (Option 1)

1) Lie on your side and position a small or large ball between your legs, just above your knees. Bend your top leg and keep your bottom leg straight.

2) Pinch your knees together and bend your bottom leg while straightening your top leg. The idea is to scissor your legs back and forth for at least 2 minutes. You can also contract by flexing your quads and then pinching your knees together with more force as you relax the muscles. Note: There's more pressure being placed on your bottom leg, so it's good to alternate after about a minute.

Knee Scissor Stack and Smash (Option 2)

Although you're targeting three areas, this option is particularly effective for mobilizing the outside of your knee. With a ball positioned between your legs, you can balance your bottom leg over a peanut tool, adding significant pressure. Again, the idea is to scissor back and forth—flexing one leg and extending the other—until you experience change.

Gap and Smash

**2 minutes
on each side**

The previous mobilizations focus on the tissues above and along the side of your knee. This one targets the area behind your knee, where your hamstring and calf cross the joint. As with the previous technique, the Gap and Smash is a great two-for-one mobilization because you're able to hit two muscle groups simultaneously. So, in addition to alleviating knee pain and tension, this technique is great for addressing tight calves and hamstrings.

The key is to work both the inside and the outside of your knee. For example, if you're spending two minutes total per knee, spend one minute working the inside and one minute working the outside.

1) Position a small ball behind your knee on the outside of your leg.

2) Sandwich the ball by scooting your butt toward your heel and compressing your knee around the ball. You can also curl your heel toward your butt and use both hands to pull your leg in tight. This creates a large compression force that targets your lower hamstring and upper calf. From here, you can implement one of two methods: floss by moving your foot around in all directions, or contract by flexing your calf and upper leg and then relax by compressing the ball deeper into your knee.

3) After about a minute of smashing the outside of your leg, reposition the ball behind your knee on the inside of your leg. Then employ the smash and floss and contract and relax methods.

PRESCRIPTION 12
Lower Leg (Calf and Shin)

This prescription can be used to treat these symptoms and restrictions:

- Ankle pain
- Calf stiffness
- Collapsed arches
- Knee pain
- Plantar fasciitis
- Shin splints

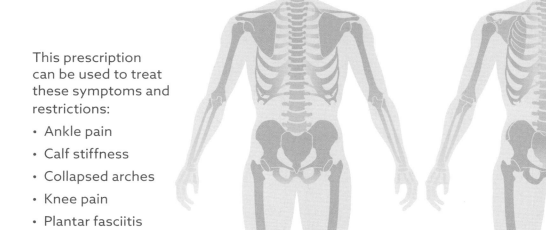

Methods:
- Contract and relax
- Pressure wave
- Smash and floss
- Tack and twist

Tools:
- Roller
- Small ball
- Large ball

Total time:
12 to 18 minutes

Overview

We've encouraged you to increase the number of steps you take as a way of promoting activity and movement. We've also emphasized the importance of standing rather than sitting at work. Well, all that time on your feet is bound to highlight how stiff your feet and calves are.

If you are plagued with knee or ankle pain, you're struggling to get your feet into good position, or your lower legs (your shins or the backs of your legs) hurt, this is the prescription to employ.

Calf Smash

The calf can be an especially sensitive area to smash. Even the toughest among us will squirm when smashing a hotspot. For this reason, we've provided several options for this mobilization, ranging from the mildly uncomfortable to the wildly nasty. Depending on your level of stiffness and your pain tolerance, you may have to start with the foam roller option and work your way up.

Before you tackle these smashes, remember that your calf muscles run from the back of your knee to your ankle. A lot of people get stuck mobilizing only one area, usually near the ankle. Don't neglect your upper calves. Targeting the entire region feeds slack to your knee and ankle joints. Like mobilizing your quads, you may have to break it up into chunks. For example, you can mobilize your upper calves in one session and then your lower calves in the next session.

**2 minutes
on each side**

Roller Calf Smash (Option 1)

Position your calf over a roller, then create a pressure wave by rolling your leg from side to side. You can also contract and relax by flexing and relaxing your calf. Remember, the target area is from your knee to your ankle. To add pressure, cross your opposite leg over your shin and lean forward.

Barbell/Hard Roller Calf Smash (Option 2)

1) For this option, Kelly is using a MobilityWOD Stick, which is essentially a piece of a barbell. The idea is to find something hard, like a rolling pin, wine bottle, or piece of pipe, over which you can rest your calf. If you have a chair with a hard backrest, you can even use that. To perform this technique, position your heel cord (Achilles tendon region) or calf on the roller, then roll your foot from side to side, point and flex your foot, and contract and relax your calf over any hotspots.

2) You can also twist the roller with your leg resting on it for a pressure wave and tack and twist effect. To increase the pressure, cross your opposite leg over your shin and lean forward.

Ball Calf Smash (Option 3)

Deskbound Variation

You can also use a large ball to smash your calves. We like to use a roller to smash our heel cords and a large ball to smash the meat of our calves. The same rules apply: pressure wave by rolling your leg from side to side, floss by moving your foot around, contract and relax by flexing your calf, and add pressure by crossing your opposite leg over your shin.

Bone Saw Calf Smash (Option 4)

1) Kneel on the ground.

2) Cross one leg over the other, positioning the shin or instep of your top leg across the calf of your bottom leg.

3) When you locate a tight area, slowly pressure wave your shin across your calf by seesawing back and forth over the muscle.

4) To add a compression force, sit your butt back and shift your body weight over your top leg. The more you sit back, the more aggressive the pressure. From here, you can continue to seesaw back and forth across your calf or employ the smash and floss or contract and relax method.

5) You can also perform this mobilization on a large pillow or place the instep of your bottom leg over a roller.

**2 minutes
on each side**

Classic Calf Stretch (Test and Retest)

We hope you have some kind of foot rail or slant board at your standing desk (see page 145) so that you can mobilize your calves throughout the day. If you do, you can use it to test and retest the effectiveness of the Calf Smash. (Stretch your calf, then do it again after you've completed one of the smashing options.) You'll find that you have a little more range of motion after completing the mobilization. If you don't have a slant board or you don't stretch your calves throughout the day, we recommend adding this basic stretch to the prescription. If you don't have a slant board, place the ball of your foot as high up on a wall, stair, or curb as possible while keeping your heel in contact with the ground.

Medial Shin Smash

**2 minutes
on each side**

The muscles that run along the inside of your leg play a crucial role in supporting the arch of your foot. If you have flat fleet (that is, your arches are collapsed), this is a great mobilization for restoring one of the key structures that has been over-stretched and misused.

We recommend this mobilization to all of our runners and clients who present with foot and ankle–related problems.

1) Sit on the ground and pin a small ball on the inside of your shin.

If mobilizing from the ground is uncomfortable or not possible, you can do this while sitting in a chair by crossing your ankle over your knee.

2) Apply downward pressure with both hands, then point your toes for a flossing effect.

3) Still applying downward pressure, flex your foot. The idea is to move your foot in various directions. You can also contract and relax, tack and twist, and pressure wave. Remember, the target area stretches from the base of your knee down to your ankle.

Stack and Smash Option

To get a bigger bang for your buck, position a second ball between your leg and the floor. Try to align the top ball over the bottom ball, then apply downward pressure. From here, employ all the smashing methods: smash and floss, contract and relax, tack and twist, and pressure wave.

Lateral Shin Smash

**2 minutes
on each side**

When it comes to mobilizing the lower leg, people tend to focus on the calf because most of the meat is on the back of the leg. But if you do a lot of standing and walking—which you should—the muscles that run down the outsides of your shinbones are bound to need attention as well. If you suffer from shin splints, for example, then ditch the flip-flops, address your standing and walking mechanics, and, for the love of your body, get to work on mobilizing the muscles running down your shins.

Target Area:
The muscles that run along the outside of your shin, from your knee down to your ankle

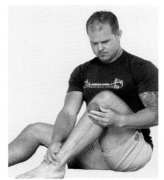

Lateral Shin Smash (Option 1)

Sit on the ground, positioning a small ball on the outside of your shin. To add a compression force, press down on your leg (aligning your hand directly over the ball) and shift your body weight over your leg. From here, you can floss by moving your foot around in all directions, contract and relax, or pressure wave by rolling back and forth over the muscles that run along the front and outside of your lower leg.

Lateral Shin Smash (Option 2)

1) Kneel on the ground and position a small ball underneath your leg on the outside of your shin. To add pressure, you can sit your butt back, drive your knee toward the floor, or reposition your center of mass over the ball.

2) Work across the muscle running along the outside of your shin, pressure waving back and forth. If you find a hotspot, stop and floss by moving your foot around in all directions (you may have to elevate your hips to do it) or contract and relax.

Classic Shin Stretch (Test and Retest)

Like the Classic Calf Stretch, this is a nice way to test the effectiveness of the Medial Shin Smash. Do this stretch before you implement one of the smashing options, noting how you feel and how far you can lift your knee off the floor, then do it again after you've completed 2 minutes of smashing. This is also a good mobilization to include in the prescription.

**2 minutes
on each side**

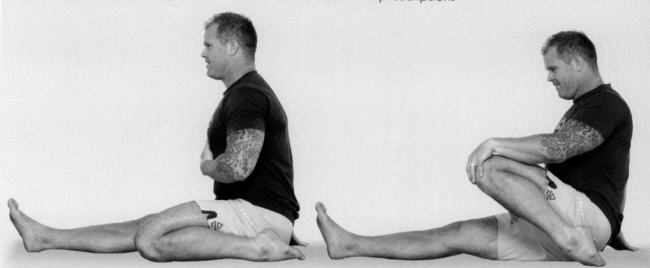

1) Sit on the ground with one leg straight out in front of you and the other leg folded behind you.

2) Lean back and lift your bent leg off the ground, keeping your toes in contact with the floor.

PRESCRIPTION 13
Ankle, Foot, and Toes

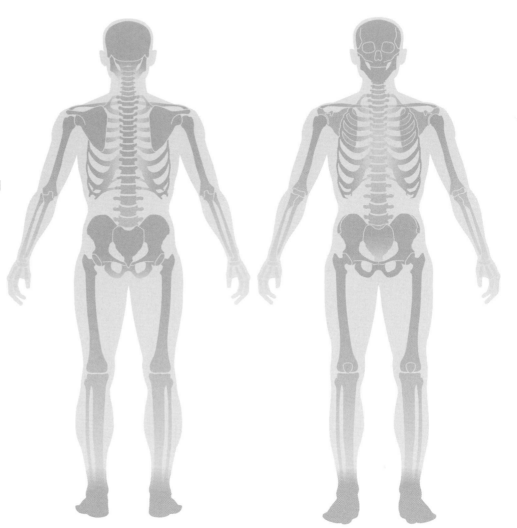

This prescription can be used to treat these symptoms and restrictions:

- Ankle pain
- Ankle range of motion (dorsiflexion, plantar flexion, plantar extension)
- Bunions
- Collapsed arches
- Flat feet
- Foot and toe stiffness
- Plantar fasciitis (foot pain)
- Turf toe

Methods:
- Contract and relax
- Pressure wave
- Smash and floss
- Tack and twist

Tools:
- Small ball

Total time:
12 minutes

Overview

When we see someone in our physical therapy practice who is suffering from foot pain—whether it is non-specific foot pain or something like plantar fasciitis—this is the mobility prescription that we recommend. This prescription serves as a blueprint for reducing soreness and relieving pain symptoms as well as keeping your feet supple and healthy.

If your feet are stiff, it is difficult to create dynamic, stable arches. Your feet endure a lot of abuse, so you need to take care of them. Ask anyone who suffers from turf toe or bunions what it's like to walk, and they will tell you that it's miserable. The good news is that you can improve the quality of your feet in a relatively short period. Make this prescription a part of your daily routine if you have flat feet (collapsed arches), bunions, plantar fasciitis, or turf toe.

Plantar Surface Smash

The plantar fascia is a big sheet of connective tissue that runs along the bottom of the foot, from the ball to the heel.

This is one of our favorite mobilizations for deskbound workers because it can be done at a desk while talking on the phone, texting, or answering emails. In other words, you can work on resolving pain and improving performance while getting work done at the same time.

**2 minutes
on each side**

Plantar Surface Smash (Option 1)

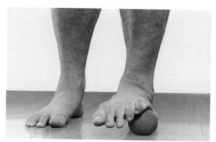 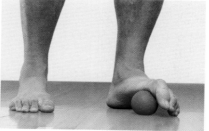

1) Step on a small ball, positioning it anywhere on your plantar surface, from your heel to your toes. Get as much weight as you can handle over the ball.

2) Strum across the arch of your foot by slowly pressure waving from side to side. When you get to the inside of your foot, you can contract and relax by curling your big toe over the ball. The key is to take your time and put in some quality work.

Forefoot Smash (Option 2)

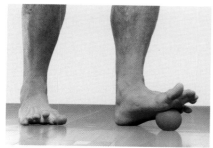 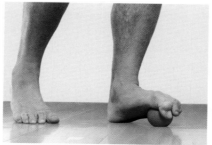 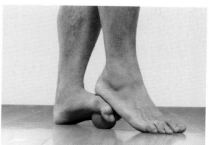

1) Step on a small ball, positioning it in the center of your forefoot.

2) Curl your toes around the ball.

3) To get more weight over your foot, step your opposite heel over your big toe and index toe and apply pressure.

Ankle Tack and Twist

**2 minutes
on each side**

In Section 6, we explained the concept of sliding surfaces—that is, how your skin, nerves, muscles, and tendons should slide and glide over one another unrestricted. When you are hypo-hydrated and sedentary, you move poorly, and you don't perform daily maintenance, these tissues can get stiff and stick together. This is what we refer to as "sliding surface dysfunction." For example, if you bend or flex your foot, the skin should slide easily over your ankle bone and tendons. If the skin doesn't slide, recognize it as a problem and work on restoring sliding surfaces to that tacked-down tissue.

That is exactly what this technique aims to do. You simply pin a ball inside, outside, and around your heel cord and employ the tack and twist method. Then, with the skin tacked down, you slide the ball around, stretching the skin. This frees the skin from the underlying surfaces and restores sliding surface function. We've seen people make dramatic improvements in their foot mobility with just a few minutes of this "scraping." It's like removing a cast from your foot.

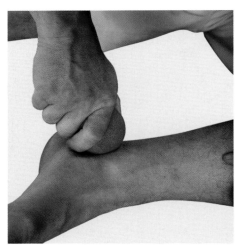

1) Sit on the floor and position a small ball on the inside of your ankle.

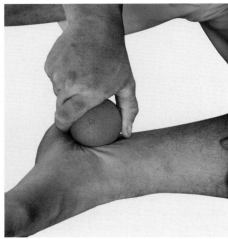

2) Applying downward pressure, twist the ball into your skin.

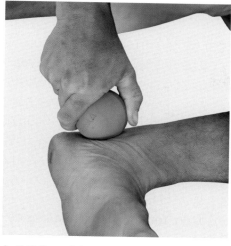

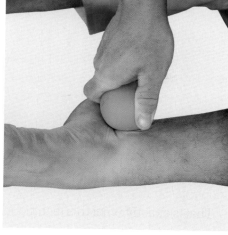

3-4) Still applying pressure, floss around the tacked-down tissue by flexing and extending your foot. The idea is to tack the skin down and then push or slide the ball in every direction. You can also whack it with your opposite hand. Do so until the skin starts to slide smoothly over the underlying surfaces.

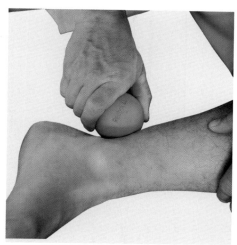

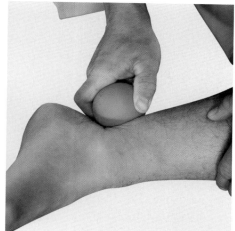

5-7) Repeat this process on your heel cord, your ankle bone, and any area around your ankle that is tacked down.

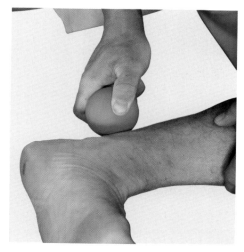

Toe Re-animator Mobilizations

**2 minutes
on each side**

Your toes are responsible for providing balance when standing and walking. They are particularly susceptible to abuse and deformation, yet we'd be willing to bet that you've neglected them your entire life. If you're like most people, you have been encasing your toes in restrictive shoes for decades. We hope that you'll do what you can to reverse this process by ditching those restrictive shoes, spending more time barefoot, and following this prescription. A good habit to get into is to perform this mobilization anytime you're hanging out on the floor watching TV. You can also do it in a chair by crossing the foot that you plan to mobilize over your opposite knee.

Finger Splice

Spread and Pull

To begin, intertwine your fingers and toes by inserting your index finger between your big toe and index toe, your middle finger between your index toe and middle toe, your ring finger between your middle toe and fourth toe, and your pinky between your fourth toe and pinky toe. You can intertwine your left hand over the top of your left foot, as shown, or intertwine your right hand around the bottom of your foot. Unless you have a structural deformation or giant fingers, you should be able to fit your fingers between your toes without discomfort. If it's painful, it's a good indication that your fascia is locked up. With your fingers and toes interlocked, twist your mid-foot joint back and forth several times. Notice that Kelly is using his opposite hand to assist with the twisting. You can also contract against the twist, then relax to go a bit farther.

As an extension of the Finger Splice, you can spread your toes by pulling them in different directions. The idea is to pull each toe up and down and from side to side. Be sure to target every pair of toes, starting with your big toe and index toe, then moving on to your index toe and middle toe, and so on.

PRESCRIPTION 14
Deskbound Rx

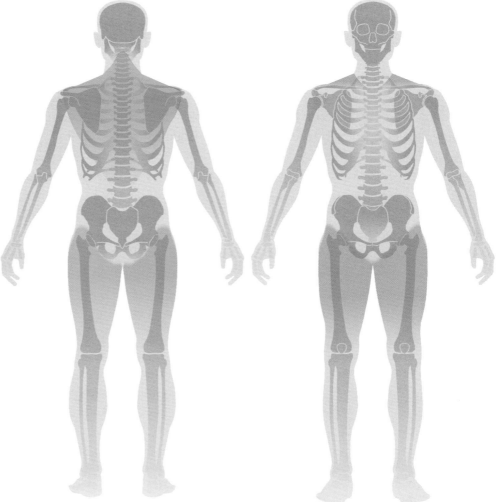

This prescription can be used to:

- Minimize the effects of being deskbound
- Open up your hips
- Resolve upper back and neck stiffness

Methods:
- Banded flossing
- Contract and relax
- Pressure wave
- Smash and floss

Tools:
- Roller
- Band
- Small ball

Total time:
14 minutes

Overview

At the beginning of this section, we explained the importance of designing personalized mobility prescriptions based on your problems of the day, your pain symptoms, and your joint and tissue restrictions. Our intention in providing mobility prescriptions that cover every region of the body is twofold:

· To get you to mobilize your entire body

· To expose you to a broad spectrum of mobility techniques so that you can figure out what works best for you

If you've performed the first 13 prescriptions, you should have a pretty good idea of where your problem areas are and how best to treat and resolve restrictions and pain. If you're limited in the squat shape, for example, you should probably do something to improve that position. If you took a road trip or had to travel for work, employing mobilizations that open up your hips is a good idea. While you can certainly continue using the prescriptions provided, we encourage you to design your own sequences using the techniques outlined in this section, or even come up with your own unique variations.

That said, we know that some of you will still want an all-encompassing prescription. Not wanting to disappoint, we've designed a general daily deskbound prescription. If we had to choose four deskbound-specific mobilizations, we would choose these. Most people will benefit from doing the T-Spine Smash, opening up their hips with the Couch Stretch, addressing hip capsule and posterior chain restrictions with the Banded Hamstring Floss, and combating neck and shoulder tension with the Trap and First Rib Smash.

There is no one-size-fits-all daily prescription that works for everyone. This is just one example of how you can cherry-pick from the techniques to create your own prescription.

T-Spine Smash
(page 261)

2 minutes

1) Sit down on the floor and position a roller behind your back at the base of your rib cage.

2) Wrap your arms across your chest into a big hug. This takes up the soft tissue slack in your upper back and pulls the scapulae out of the way, allowing you to target your thoracic spine.

3) Keeping the roller in the same spot, arch back over the roller. In this position, take a big breath and try to arch farther over the roller as you exhale. You can also elevate your hips to increase the pressure.

T-Spine Smash: Side-to-Side Variation

4) With your arms wrapped tightly around your body, sit up as if you were doing a crunch.

5) Keeping the majority of your body weight over the roller, twist from your hips or rotate your entire upper body. From here, you can arch or side-bend over the roller or roll up and down the side of your back.

6) You can also roll back and forth, seesawing from side to side.

T-Spine Smash: Overhead Variation

7) Return to the start position. Raising your arms overhead can cause you to overextend at the lumbar spine, so get your abs tight before moving on to the next step.

8) With your abs braced, raise your arms overhead, interlocking your thumbs. Keep your elbows locked out and reach for the ceiling.

9) Now arch back over the roller.

10) To add more pressure, drive your heels into the ground and elevate your hips.

11) Still arching back, lower your butt to the floor.

Couch Stretch
(pages 307 and 308)

**2 minutes
on each side**

1) On your hands and knees, back your feet up to a wall.

2) Slide one leg back, driving your knee into the corner and positioning your shin and foot flush with the wall.

3) Post up on your opposite leg, keeping the shin of that leg as vertical as possible.

4) Squeeze the glute of your rear leg. If you're unable to post up on your lead leg because you're too stiff, position a small box or chair in front of you for added stability.

5) After hanging out in the previous position for a minute or longer, raise your torso into the upright position. If you find it difficult to support the weight of your body in the upright position, place a box or chair in front of you for added stability.

Banded Hamstring Floss
(page 318)

1) Hook your leg through a band, wrapping it around your hip. Then create tension by walking forward and hinging forward from the waist. Assume a sprinter's stance with your free leg slightly front and your banded leg bent. If you can't reach the ground without rounding your back, position a chair or stool in front of you.

2) Keeping your back as flat as possible, floss by straightening your banded leg and driving your hips back.

Trap and First Rib Smash
(page 265)

1) Nestle a ball between your clavicle and the base of your trap and neck.

2) Pinning the ball in place against a wall, drive your weight into the ball. From here, contract by engaging your trap and shoulder, then relax, pushing the ball deeper into your neck and shoulder. After a few contractions, raise your arm overhead and move your hand behind your back. You can also use your hand to gently pull your head away from the ball. Keeping pressure into the ball, you can lower your upper body and allow the ball to roll over the top of your trap.

Afterword

Being deskbound doesn't mean that you're doomed to a life of pain and accelerated decline. There's a simple solution, and we hope that this book will serve as your road map to a better quality of life. But the decision is yours to make. When it comes to your body, you have options and control. You can decide, for example, to reduce optional sitting in your life, focus on your mechanics, set reminders for movement breaks, and mobilize for at least 10 minutes every day. As adults, we have the power of choice. Kids, however, do not.

Think about it: would you stick your own child (or any child, for that matter) in a smoky bar for nine hours? We certainly hope not. Yet this is the equivalent of what we're doing every time we allow our children to sit for 10 to 14 hours a day. We have to take responsibility and treat this for what it is: a public health crisis. Sitting and inactivity are predisposing our youth to pain and illness and a reduced quality of life. To solve this problem, we not only need to pass along and teach the principles, guidelines, and techniques

outlined in this book, but also need to remove the innocuous environmental load of sitting in the place where they spend the majority of their waking hours—school.

StandUp Kids

In June 2014, we volunteered at field day at our daughters' school. We always volunteer at the sack races, because of all the events, we think the sack race is the most interesting and the most athletic. What we saw was alarming. The kids—most of them not obese—lacked the range of motion to lift their legs to get into the sack. Many of them had difficulty lifting their knees to their chests in order to put their feet in the bag. Then, when they would jump, they had such insufficient hip range of motion that they couldn't get into full extension. That means they couldn't jump well or efficiently. It blew our minds, and it scared us.

Most of these children were visibly healthy. The only thing we could conclude is that these children had lost critical hip range of motion. And these are first through fifth graders! It was obvious to us: this was the result of sitting. The only environmental load that could cause something like this in young kids was sitting too much. That was the first moment—our first spark of realization about what sitting was doing to children. We were actually embarrassed that we'd be counseling all of our clients to move to standing

desks and mindlessly sending our kids to school to sit all day. We realized that we had to take action.

We dove into the research and found Dr. Mark Benden at Texas A&M University. His comprehensive two-year study of standing desks involved 500 elementary school students. It measured changes in the students' calorie expenditure and classroom engagement. His study showed that normal-weight kids burned between 15 and 25 percent more calories at a standing desk. For obese students, the impact was even greater. They burned between 25 and 35 percent more calories while standing. The research also showed that kids at standing desks are 12 percent more engaged, which translates to 7 minutes per hour, 45 minutes per day, and 135 hours per year of *more* engagement.

We realized that standing not only would help prevent a myriad of orthopedic problems in children, but also could be a simple and elegant tool in combating childhood obesity, increasing classroom concentration, and much more.

We were very motivated to make a change at our own children's school. We spoke to Principal Tracy Smith. We assumed that we'd need to make a big, detailed presentation, make our case over the course of many months, and make our case to the district and the school board. To our surprise and delight, within five minutes of our "pitch," Principal Smith said, "I'm in!" We found kids' standing desks online and were attracted to them because they were equipped with a swinging foot bar. We bought 25 and installed them in our older daughter's fourth grade classroom in August 2014. We were a go: these kids were no longer sitting all day at school. The program was so successful that in January 2015 we bought desks for the school's two other fourth grade classes and one first-grade class, and 100 kids were standing. By April 2015, we had launched StandUp Kids in partnership with Donors Choose, the innovative nonprofit that supports public school teachers. The mission of StandUp Kids is to get every public school child at a standing desk within 10 years, to combat the epidemic of sedentary lifestyles and inactivity, and to better reflect 21st-century education goals.

We then launched an ambitious crowdfunding campaign to raise funds to purchase 350 more desks to the tune of $110,000. Thanks in large part to our champion, Tim Ferriss, along with the most individual donors ever to donate to a Donors Choose project (nearly 900 people), we raised the money in less than two months. Vallecito Elementary School in San Rafael, California, is now the first all-standing school in the world.

Since that time, the research has continued to pour in supporting kids using standing desks at school. For example, a recent study showed that the continued use of standing desks at school is associated with significant improvements in executive function and working memory capabilities. The researchers concluded that standing desks may provide advantages similar to those of exercise programs, which boost brain activity by enhancing blood flow. In addition, new research from the Netherlands published in the journal *Diabetologia* links sitting for as little as 40 minutes with a significantly increased risk of diabetes. According to *Education World*, this is just one of many studies making a strong case for mandating standing desks in schools.

The evidence is in that standing at school is better for kids' bodies and minds. As of the time of publication of this book, we estimate that StandUp Kids has helped more than 15,000 students across the United States successfully transition to standing desks, 75 percent of whom attend low-income schools. To learn more and to support a standing classroom near you, visit www.StandUpKids.org.

Beyond *Deskbound*

There is a lot more to our system than what you're holding in your hands. To help with your continuing education, we've compiled a list of additional resources. And to help you follow the guidelines, principles, and techniques outlined in this book, we've provided a list of products and tools.

Websites

MobilityWOD.com—In 2010, we started MobilityWOD.com, an instructional-based website designed to help people resolve pain, prevent injury, and optimize athletic performance. We invite you to visit us and hope that you will join our knowledgeable and supportive community.

RogueFitness.com—Offering a ton of mobility and exercise equipment, Rogue has everything you need to upgrade your mobility home gym.

StandUpKids.org—This organization provides education on the health risks associated with a sedentary lifestyle and resources on why standing desks create an environment in which children can be healthier and more active. The mission of StandUp Kids is to get every public school child at standing desk within 10 years.

Videos

On our website, MobilityWOD.com, there are more than 1,600 videos and counting, covering everything from squatting, standing, and sitting mechanics to treating low back, neck, and shoulder pain to breathing and improving sleep. We film daily mobility prescriptions and have hundreds of archived videos. To preview some of our videos, go to www.MobilityWOD.com/preview/

We also encourage you to watch the "Deskbound" talk that Kelly did for Google, which helped shape the foundation for this book: www.youtube.com/watch?v=kfg_e6YG37U

Books

Becoming a Supple Leopard: The Ultimate Guide to Resolving Pain, Preventing Injury, and Optimizing Athletic Performance by Dr. Kelly Starrett with Glen Cordoza (Victory Belt Publishing, 2013). Written primarily with coaches and athletes in mind, this *New York Times* and *Wall Street Journal* bestseller unveils Kelly's system for movement and mobility and includes all the exercise movements and mobility techniques that comprise his system. It's the bedrock underlying the principles and concepts covered in *Deskbound*.

Ready to Run: Unlocking Your Potential to Run Naturally by Dr. Kelly Starrett with T. J. Murphy (Victory Belt Publishing, 2014). Another *New York Times* bestseller offering 12 standards that will prepare your body for a lifetime of top-performance running.

The Chair: Rethinking Culture, Body, and Design by Galen Cranz (W. W. Norton & Company, 1998). This book is a call to action. Although Cranz covers the fascinating history and evolution of the chair, she challenges traditional ergonomic theory and provides compelling evidence linking back pain to conventional seating.

Could You Stand to Lose? Weight Loss Secrets For Office Workers by Mark E. Benden, PhD, CPE (Trinity River Publishing, 2008). Another excellent book detailing why sitting is bad and how standing is a healthier alternative. Benden offers practical advice for changing basic work habits to help you lose weight. He also outlines what equipment is best for your workplace and even offers strategies for getting your company on board with standing desks.

Get Up!: Why Your Chair Is Killing You and What You Can Do About It by James A. Levine, MD (Palgrave Macmillan, 2014). This book explains the science behind the perils of sitting and offers solutions for correcting the problem. Dr. Levine is an innovator and the godfather of the stand-up revolution. We highly recommend his book. It investigates the scientific concepts presented in *Deskbound* in much greater detail.

The Power of Habit: Why We Do What We Do in Life and Business by Charles Duhigg (Random House, 2011). To get the most out of *Deskbound*, you need to create new movement, mechanics, and mobility habits. And to do that, you need to understand how habits work. In this book, Duhigg explains why habits exist and how they can be changed.

Spark: The Revolutionary New Science of Exercise and the Brain by John J. Ratey, MD, with Eric Hagerman. Wondering how movement, or the lack thereof, can affect brain health and performance? Ratey provides the answers. He explores the link between exercise and the brain by explaining how aerobic exercise or activity physically remodels our brains for peak performance.

Products

Mobility Tools

Battlestar, Gemini, and Supernova mobility tools (www.MobilityWOD.com/product-category/gear)

Rogue Monster Bands (RogueFitness.com)

Standing Desks

The desk pictured throughout the book is the Jarvis desk, available at www.ergodepot.com.

IKEA also sells an affordable standing desk called the Bekant.

Sit to Stand Converters

Oristand (http://oristand.co)

Varidesk (www.varidesk.com)

Foot Rests

Fluidstance Level (www.fluidstance.com)

Rogue Fidget Bar (www.MobilityWOD.com/product-category/gear)

Blue Light Blockers

Blue light–blocking products (www.lowbluelights.com)

Gunnar eyewear (www.gunnars.com)

Uvex eyewear (www.uvex.us)

Compression Socks and Tights

Reebok (www.reebok.com/us/compression)

SKINS (www.skins.net/usa/)

Helpful Apps

Break Reminders and Focus

Focus Time (iOS) (http://focustimeapp.com/)

Marinara Timer (Web) (www.marinaratimer.com)

Stand Up! (iOS) (www.raisedsquare.com/standup/)

Time Out (Mac) (www.dejal.com/timeout/)

Tomighty (Mac/Windows) (www.tomighty.org)

Eye Health

EyeLeo (Windows) (http://eyeleo.com/)

f.lux (Mac/Windows) (https://justgetflux.com/)

Twilight (Android) (http://twilight.urbandroid.org/)

Notes

Introduction

1. Mary MacVean, "'Get Up!' or Lose Hours of Your Life Every Day, Scientist Says," *Los Angeles Times*, January 24, 2014, www.latimes.com/science/sciencenow/la-sci-sn-get-up-20140731-story.html.

2. James A. Levine, *Get Up! Why Your Chair Is Killing You and What You Can Do About It* (New York: St. Martin's Press, 2014): 70–71.

3. Aviroop Biswas, Paul I. Oh, Guy E. Faulkner, Ravi R. Bajaj, Michael A. Silver, Marc S. Mitchell, and David A. Alter, "Sedentary Time and Its Association with Risk for Disease Incidence, Mortality, and Hospitalization in Adults: A Systematic Review and Meta-Analysis," *Annals of Internal Medicine* 162, no. 2 (2015): 123–132.

4. J. Lennert Veerman, Genevieve N. Healy, Linda J. Cobiac, Theo Vos, Elisabeth A. H. Winkler, Neville Owen, and David W. Dunstan, "Television Viewing Time and Reduced Life Expectancy: A Life Table Analysis," *British Journal of Sports Medicine* 46 (2012): 927–930.

5. Mary Shaw, Richard Mitchell, and Danny Dorling, "Time for a Smoke? One Cigarette Reduces Your Life by 11 Minutes," *BMJ* 320, no. 7226 (2000): 53.

6. MacVean, "'Get Up!' or Lose Hours of Your Life Every Day, Scientist Says."

7. Eric Jensen, "Moving with the Brain in Mind," *Educational Leadership* 58, no. 3 (2000): 34–37.

8. "Health Topics: Physical Activity," World Health Organization, www.who.int/topics/physical_activity/en/.

9. Levine, *Get Up!*, 103.

10. U.S. Department of Health and Human Services, *Physical Activity and Health: A Report of the Surgeon General* (Atlanta, GA: U.S. Department of Health and Human Services, Centers for Disease Control and Prevention, National Center for Chronic Disease Prevention and Health Promotion, 1996), www.cdc.gov/nccdphp/sgr/index.htm.

11. U.S. Department of Health and Human Services, *The Power of Prevention: Chronic Disease . . . The Public Health Challenge of the 21st Century* (Atlanta, GA: U.S. Department of Health and Human Services, Centers for Disease Control and Prevention, National Center for Chronic Disease Prevention and Health Promotion, 2009), www.cdc.gov/chronicdisease/pdf/2009-Power-of-Prevention.pdf.

12. "Back Pain," MedLine Plus, National Institutes of Health, www.nlm.nih.gov/medlineplus/backpain.html; Pat Anson, "Lower Back Pain Is #1 Cause of Disability," National Pain Foundation, www.thenationalpainfoundation.org/Lower-Back-Pain-is-Number-1-Cause-of-Disability.

13. M. Mehra, K. Hill, D. Nicholl, and J. Schadrack, "The Burden of Chronic Low Back Pain with and without a Neuropathic Component: A Healthcare Resource Use and Cost Analysis," *Journal of Medical Economics* 15, no. 12 (2012): 245–252.

14. U.S. Department of Labor, Occupational Heath & Safety Administration, "Preventing Repetitive Stress Injuries," December 10, 1996, www.osha.gov/pls/oshaweb/owadisp. show_document?p_table=SPEECHES&p_id=206.

15. Alpa V. Patel, Leslie Bernstein, Anusila Deka, Heather Spencer Feigelson, Peter T. Campbell, Susan M. Gapstur, Graham A. Colditz, and Michael J. Thun, "Leisure Time Spent Sitting in Relation to Total Mortality in a Prospective Cohort of US Adults," *American Journal of Epidemiology*, 172, no. 4 (2010): 419–429.

16. Teresa Watanabe, "Just 31% of California Students Pass P.E. Test," *Los Angeles Times*, December 3, 2011, http://articles.latimes.com/2011/dec/03/local/ la-me-fitness-schools-20111203.

17. "2008 Physical Activity Guidelines for Americans Summary," U.S. Department of Health and Human Services, Office of Disease Prevention and Health Promotion, http:// health.gov/paguidelines/guidelines/summary.aspx.

18. Mary Story, Marilyn S. Nanney, and Marlene B. Schwartz, "Schools and Obesity Prevention: Creating School Environments and Policies to Promote Healthy Eating and Physical Activity," *Milbank Quarterly* 87, no. 1 (2009): 71–100.

19. "Generation M2: Media in the Lives of 8- to 18-Year-Olds," Henry J. Kaiser Family Foundation, January 20, 2010, http://kff.org/other/event/ generation-m2-media-in-the-lives-of/.

20. "Average Number of Hours in the School Day and Average Number of Days in the School Year for Public Schools, by State: 2007–08," National Center for Education Statistics, https://nces.ed.gov/surveys/sass/tables/sass0708_035_sls.asp.

21. Tala H. I. Fakhouri, Jeffrey P. Hughes, Vicki L. Burt, MinKyoung Song, Janet E. Fulton, and Cynthia L. Ogden, "Physical Activity in U.S. Youth Aged 12–15 Years, 2012," National Center for Health Statistics Data Brief No. 141, 2014, www.cdc.gov/nchs/data/databriefs/ db141.htm.

22. M. C. McDonald, "Active Transport to School: Trends Among U.S. Schoolchildren, 1969–2001," *American Journal of Preventive Medicine* 32, no. 6 (2007): 509–516.

23. Committee on Physical Activity and Physical Education in the School Environment, Food and Nutrition Board, Institute of Medicine, "Status and Trends of Physical Activity Behaviors and Related School Policies," in *Educating the Student Body: Taking Physical Activity and Physical Education to School*, ed. H. W. Kohl III and H. D. Cook (Washington, DC: National Academies Press, 2013), www.ncbi.nlm.nih.gov/books/ NBK201496/.

24. "TV Basics: TV Sets Per Household," Television Bureau of Advertising, http:// archivesite.tvb.org/rcentral/mediatrendstrack/tvbasics/07_5_TV_Per_HH.asp.

25. "The Epidemic of Childhood Obesity: Learn the Facts," Let's Move!, www.letsmove. gov/learn-facts/epidemic-childhood-obesity.

26. Lorrene D. Ritchie, Susan L. Ivey, Maggie Masch, Gail Woodward-Lopez, Joanne Ikeda, and Pat Crawford, *Pediatric Overweight: A Review of the Literature* (Berkeley, CA: Center for Weight and Health, College of Natural Resources, UC Berkeley, 2001).

27. Nicholas Staropoli, "A Reader Asks: Is Life Expectancy in America Declining?," American Council on Science and Health, June 3, 2015, http://acsh. org/2015/06/a-reader-asks-is-life-expectancy-in-america-declining/.

28. Biswas et al., "Sedentary Time and Its Association with Risk for Disease Incidence, Mortality, and Hospitalization in Adults."

29. Ranjana K. Mehta, Ashley E. Shortz, and Mark E. Benden, "Standing Up for Learning: A Pilot Investigation on the Neurocognitive Benefits of Stand-Biased School Desks," *International Journal of Environmental Research and Public Health* 13, no. 1 (2016): 59.

30. Biswas et al., "Sedentary Time and Its Association with Risk for Disease Incidence, Mortality, and Hospitalization in Adults."

31. "The Shocking 'Text Neck' X-Rays That Show How Children as Young as SEVEN Are Becoming Hunch Backs Because of Their Addiction to Smart Phones," *Daily Mail Australia*, October 15, 2015, www.dailymail.co.uk/news/article-3274835/Shocking-X-rays-teenagers-text-neck.html.

32. Levine, *Get Up!*, 70–71; Daniela Schmid and Graham Colditz, "Sedentary Behavior Increases the Risk of Certain Cancers," *Journal of the National Cancer Institute* 106, no. 7 (2014).

33. J. A. Bell, M. Hamer, G. D. Batty, A. Singh-Manoux, S. Sabia, M. Kivimaki, "Combined Effect of Physical Activity and Leisure Time Sitting on Long-Term Risk of Incident Obesity and Metabolic Risk Factor Clustering," *Diabetologia* 57, no. 10 (2014): 2048–2056.

34. Levine, *Get Up!*, 24–27.

35. James A. Levine, Mark W. Vander Weg, James O. Hill, and Robert C. Klesges, "Non-Exercise Activity Thermogenesis: The Crouching Tiger Hidden Dragon of Societal Weight Gain," *Arteriosclerosis, Thrombosis, and Vascular Biology* 26, no. 4 (2006): 729–736.

36. Levine, *Get Up!*, 29–32.

37. Christopher Berglund, "Why Does Physical Activity Drain Human Brain Power?" *Psychology Today*, December 4, 2014, www.psychologytoday.com/blog/the-athletes-way/201412/why-does-physical-inactivity-drain-human-brain-power.

38. John Ratey with Eric Hagerman, *Spark: The Revolutionary New Science of Exercise and the Brain* (New York: Little, Brown, 2008), 3–8.

39. Berglund, "Why Does Physical Activity Drain Human Brain Power?"

40. "Sit. Stand. Move. Repeat. The Importance of Moving as a Natural Part of the Workday," Herman Miller, www.hermanmiller.com/research/solution-essays/sit_stand_move_repeat.html.

41. Archana Singh-Manoux, Melvyn Hillsdon, Eric Brunner, and Michael Marmot, "Effects of Physical Activity on Cognitive Functioning in Middle Age," *American Journal of Public Health* 95, no. 12 (2005): 2252–2258.

42. Laura Chaddock, Michelle W. Voss, and Arthur F. Kramer, "Physical Activity and Fitness Effects on Cognition and Brain Health in Children and Older Adults," *Kinesiology Review* 1, no. 1 (2012): 37–45.

43. Ratey, *Spark*, 9–25.

44. Valerie Strauss, "Why So Many Kids Can't Sit Still in School Today," *Washington Post*, July 8, 2014, www.washingtonpost.com/news/answer-sheet/wp/2014/07/08/why-so-many-kids-cant-sit-still-in-school-today/; James Hamblin, "Exercise Is

ADHD Medication," *The Atlantic*, September 24, 2014, www.theatlantic.com/health/archive/2014/09/exercise-seems-to-be-beneficial-to-children/380844/.

45. Ratey, *Spark*, 245–268.

46. "Famous People with Standing Desks," Notsitting.com, http://notsitting.com/standing-desks/general-info/famous-people/.

47. Richard Branson, "Why You Should Stand Up in Meetings," *Virgin Blog*, April 6, 2015, www.virgin.com/richard-branson/why-you-should-stand-up-in-meetings.

48. "Famous People with Standing Desks," Notsitting.com, http://notsitting.com/standing-desks/general-info/famous-people/.

Section 1

1. Erik Dalton, "Forward Head Posture," Freedom from Pain Institute, http://erikdalton.com/media/published-articles/forward-head-posture/.

Section 2

1. L. A. Lipsitz, I. Nakajima, M. Gagnon, T. Hirayama, C. M. Connelly, and H. Izumo, "Muscle Strength and Fall Rates Among Residents of Japanese and American Nursing Homes: An International Cross-Cultural Study," *Journal of the American Geriatrics Society* 42, no. 9 (1994): 953–959.

Section 4

1. Bryan Walsh, "The Dangers of Sitting at Work—and Standing," *Time*, April 13, 2011, http://healthland.time.com/2011/04/13/the-dangers-of-sitting-at-work—and-standing/.

2. Mark E. Benden, *Could You Stand to Lose? Weight Loss Secrets for Office Workers*, 2nd ed. (Trinity River Publishing, 2008), 67.

3. "Computer Vision Syndrome," American Optometric Association, www.aoa.org/patients-and-public/caring-for-your-vision/protecting-your-vision/computer-vision-syndrome?sso=y.

4. Harvard Health Publications, Harvard Medical School, "Blue Light Has a Dark Side," *Harvard Health Letter*, May 1, 2012 (updated September 2, 2015), www.health.harvard.edu/staying-healthy/blue-light-has-a-dark-side.

5. Alison Griswold, "To Work Better, Just Get Up from Your Desk," *Forbes*, June 12, 2012, www.forbes.com/sites/alisongriswold/2012/06/12/to-work-better-just-get-up-from-your-desk/.

6. K. Forcier, L. R. Stroud, G. D. Papandonatos, B. Hitsman, M. Reiches, J. Krishnamoorthy, and R. Niaura, "Links between Physical Fitness and Cardiovascular Reactivity and Recovery to Psychological Stressors: A Meta-Analysis," *Health Psychology* 25, no. 6 (2006): 723–739.

7. Julia Gifford, "We Tested Standing Desks—Here's Proof They Make You More Productive," *ReadWrite*, September 26, 2013, http://readwrite.com/2013/09/26/standing-desks-productivity.

8. Andrew P. Knight and Markus Baer, "Get Up, Stand Up: The Effects of a Non-Sedentary Workspace on Information Elaboration and Group Performance," *Social Psychological and Personality Science* 5, no. 8 (2014): 910–917.

Section 5

1. Leonardo Barbosa Barreto de Brito, Djalma Rabelo Ricardo, Denise Sardinha Mendes Soares de Araújo, Plínio Santos Ramos, Jonathan Myers, and Claudio Gil Soares de Araújo, "Ability to Sit and Rise from the Floor as a Predictor of All-Cause Mortality," *European Journal of Preventive Cardiology* (2012).

2. "Sitting Straight 'Bad for Backs,'" *BBC News,* November 28, 2006, http://news.bbc.co.uk/2/hi/6187080.stm.

3. Benden, *Could You Stand to Lose?*, 40.

4. Galen Cranz, *The Chair: Rethinking Culture, Body, and Design* (New York: W. W. Norton, 2000), 104.

5. "Stand Up, Walk Around, Even Just for '20 Minutes,'" *NPR Books,* May 9, 2012, www.npr.org/2012/05/09/152336802/stand-up-walk-around-even-just-for-20-minutes.

Section 6

1. Laura Donnelly, "Sleep Deprivation 'as Bad as Smoking,'" *The Telegraph,* July 27, 2015, www.telegraph.co.uk/news/health/11765723/Sleep-deprivation-as-bad-as-smoking.html.

2. "Consequences of Insufficient Sleep," Harvard Medical School, Division of Sleep Medicine, http://healthysleep.med.harvard.edu/healthy/matters/consequences.

Afterword

M. E. Benden, J. J. Blake, M. L. Wendel, and J. C. Huber Jr., "The Impact of Stand-Biased Desks in Classrooms on Calorie Expenditure in Children," *American Journal of Public Health* 101, no. 8 (2011): 1433–1436.

M. E. Benden, H. Zhao, C. E. Jeffrey, M. L. Wendel, and J. J. Blake, "The Evaluation of the Impact of a Stand-Biased Desk on Energy Expenditure and Physical Activity for Elementary School Students," *International Journal of Environmental Research and Public Health* 11, no. 9 (2014): 9361–9375.

M. Benden et al., "The Effect of Stand-Biased Desks on Academic Engagement: An Exploratory Study," *International Journal of Health Promotion and Education* 53, no. 5 (2015): 271–280.

Ranjana K. Mehta, Ashley E. Shortz, and Mark E. Benden, "Standing Up for Learning: A Pilot Investigation on the Neurocognitive Benefits of Stand-Biased School Desks," *International Journal of Environmental Research and Public Health* 13, no. 1 (2016): 59.

Julianne D. van der Berg, Coen D. A. Stehouwer, Hans Bosma, Jeroen H. P. M. van der Velde, Paul J. B. Willems, Hans H. C. M. Savelberg, Miranda T. Schram, et al., "Associations of Total Amount and Patterns of Sedentary Behavior with Type 2 Diabetes and the Metabolic Syndrome: The Maastricht Study," *Diabetologia,* published electronically February 2, 2016, doi: 10.1007/s00125-015-3861-8.

Nicole Gorman, "Case for Standing Desks in the Classroom Grows as Further Research Links Sitting to Diabetes," *Education World,* February 3, 2016, www.educationworld.com/a_news/case-standing-desks-classroom-grows-further-research-links-sitting-diabetes-1903288572#sthash.FPAvdpi5.dpuf.

Acknowledgments

Thank you massively to the following people, without whose contributions and support this book would not have been written:

Dr. Mark Benden, Dr. James Levine, and Dr. John Ratey for their trailblazing work and extensive and thoughtful research on the ill effects of sitting too much and the positive effects of standing and moving on the brain and body. This book, along with our organization, StandUp Kids, would not exist without them and their work.

Our brother, Tom Wiscombe, for his beautiful book cover design and his continuing support, advice, and general wisdom. Our mom, Janet Wiscombe, for her editing, brainstorming, love, and never-ending support of our work and lives. Ben Lieb for thinking of the perfect subtitle to this book.

Dave Beatie, Margaret Garvey, and the entire MobilityWOD staff for their hard work, humor, and support on this book and all of our projects.

Christopher Jerard and the Inkwell team for their tireless work and dedication in making sure people know not only that this book exists, but how important it is.

Darren Miller for taking the beautiful photos and making us look way younger and more beautiful than we really are.

Ergo Depot for loaning us the standing desk pictured throughout the book.

Our publisher, Erich Krauss, and the Victory Belt staff for the opportunity to publish this and our other books and for their unflagging support of our work.

Pam Mourouzis, our incredible editor, for her drive, direction, and optimism. This book would not have made it to print without you.

Donors Choose for their support of and advocacy for StandUp Kids, which we consider to be the most important work we do. Tim Ferriss for his epic support in making the first all-standing school in the world a reality. The StandUp Kids board of directors and advisors, including Pam Lauper, Becca Russell, Nate Forester, Dr. Mark Benden, Dr. James Levine, Dr. John Ratey, Ben Greenfield, Drew Amoroso, Allison Belger, Gray Cook, Roop Sihota, John Post, Jeff and Mikki Martin, and Tracy Jerard.

Index

Index

programming, for mobility, 250–253
pronation, of ankles, 92
proprioceptive neuromuscular
    facilitation (PNF), 231
psoas, tight, 54
pulling
    about, 129
    "break the bar," 129–130
    with one hand, 132
    "screw your hands into the
        ground," 131
pushing
    about, 129
    "break the bar," 129–130
    with one hand, 132
    "screw your hands into the
        ground," 131

Q
QL Side Smash, 280–281
Quad Smash, 313–314
    break, 171
    as example of sliding surface
        mobilization, 214
    as example of smashing technique,
        231
quads, mobility prescriptions for,
    312–325

R
range of motion
    improving, 36–37, 222–229
    improving with mobility
        prescriptions, 253
    loss of, due to rounded spine, 44
    mobility prescriptions for,
        260–277, 284–339
    whole-body mobility prescriptions
        for, 255
Ratey, John J.
    on physical activity and
        movement, 23
    Spark: The Revolutionary New
        Science of Exercise and the
        Brain, 22, 351–352
Ready to Run: Unlocking Your
    Potential to Run Naturally (Starrett
    and Murphy), 13, 205, 351
Reebok (website), 353

referred pain, 221
resources, 350–353
restrictions, treating specific, 37
retesting positions, 246
Reynolds, Gretchen, The First
    Twenty Minutes, 196
Rogue Fitness (website), 350
Rogue Monster Bands, 243, 352
Rolf, Ida, 221
Rolfing, 221
roller, 239
Roller Calf Smash, 327
rotation
    about, 74–77
    illustration of, 76
    stability through the shoulders,
        77–79
rotator cuff pain, mobility
    prescriptions for, 272–277
rounded spine
    about, 40–41
    diaphragm dysfunction, 45
    fatty neck hump, 48
    headaches, 46
    illustrated, 42–43
    jaw pain (TMJD), 48
    loss of normal range of motion,
        44
    low back pain, 47
    neck pain, 46
    numbness and tingling, 45

S
sacrum, 40
scapula, mobility prescriptions for,
    260–265
sciatica, mobility prescriptions for,
    278–283, 298–303, 312–319
"screw your hands into the ground"
    cue, 131
seat belts, in airplanes, 201–202
seats
    airplane, 200–202
    car, 202–203
    of chairs, 197–198
self-maintenance, 206. See also
    basic maintenance
shallow neck breathing pattern, 67
shape, of chair seat, 198

shin splints, mobility prescriptions
    for, 326–333
shins
    mobility prescriptions for, 326–333
    verticalizing, 100–101
shoes
    for standing workstation, 138,
        141–142
    walking and, 90–91
Shoulder Extension Mobilization,
    270–271
Shoulder Internal Rotation Test, 227
Shoulder Opener break, 172, 178
Shoulder Rotator Smash and Floss,
    273–274
shoulders. See also natural body
    principles
    about, 32, 120–121
    during car travel, 203
    carrying, 122–128
    faults and corrections, 132–135
    illustrated, 122–128, 129–130, 131,
        133–135
    mobility prescriptions for, 272–277
    pulling, 129–135
    pushing, 129–135
    rotational stability through, 77–79
    stable, 33
    stable shapes for, 122–128
Side Hip Smash, 301
side slouch
    about, 56
    illustrated, 57
Single Leg Flexion and External
    Rotation, 216, 309–311
sit to stand converters, 352
sitting
    about, 179
    airplane, 200–202
    car seat, 202–203
    chair, 197–199
    children and, 14–15
    cholesterol and, 16
    diabetes and, 17
    on edge of seat, 193–194
    figure-four, 185
    golden rules of, 183
    on the ground, 184–186

About the Authors

Dr. Kelly Starrett is the author of the *New York Times* bestseller *Becoming a Supple Leopard,* which has revolutionized how coaches, athletes, and every-day humans approach movement and athletic performance. Dr. Starrett is a co-founder of San Francisco CrossFit and MobilityWOD.com, where he shares his innovative approach to movement, mechanics, and mobility with coaches and athletes. He travels around the world teaching his wildly popu-lar Movement & Mobility Course and works with elite Army, Navy, Air Force, Marines, and Coast Guard forces; athletes from the NFL, NBA, NHL, and MLB; and nationally ranked and world-ranked strength and power athletes. He consults with Olympic teams and universities and is a featured speaker at strength and conditioning conferences nationwide. Dr. Starrett's work is not limited to coaches and athletes; his methods apply equally well to children, desk jockeys, and anyone dealing with injury and chronic pain. He believes that every human being should know how to move and be able to perform basic maintenance on themselves.

Juliet Starrett is an attorney, athlete, and entrepreneur. She is co-founder and CEO of San Francisco CrossFit and MobilityWOD.com. As a mother and co-founder of the nonprofit StandUp Kids, Juliet is committed to getting every public school child at a standing desk within 10 years in order to combat the epidemic of sedentary lifestyles. In her earlier life, Juliet was a professional athlete, paddling on the U.S. Extreme Whitewater Team from 1997 to 2000. She won two world championships and five national titles.

Glen Cordoza is a *New York Times* and *Wall Street Journal* bestselling co-author of *Becoming a Supple Leopard* and a former professional mixed martial artist and Muay Thai boxer. He is one of the most published authors on the topics of MMA, Brazilian Jiu-Jitsu, Muay Thai, and general fitness, with 24 books to his credit.